Law 3.0

Putting technology front and centre in our thinking about law, this book introduces Law 3.0: the future of the legal landscape.

Technology not only disrupts the traditional idea of what it is 'to think like a lawyer,' as per Law 1.0; it presents major challenges to regulators who are reasoning in a Law 2.0 mode. As this book demonstrates, the latest developments in technology offer regulators the possibility of employing a technical fix rather than just relying on rules – thus, we are introducing Law 3.0. Law 3.0 represents, so to speak, the state we are in and the conversation that we now need to have, and this book identifies some of the key points for discussion in that conversation. Thinking like a lawyer might continue to be associated with Law 1.0, but from 2020 onward, Law 3.0 is the conversation that we all need to join. And, as this book argues, law and the evolution of legal reasoning cannot be adequately understood unless we grasp the significance of technology in shaping both legal doctrine and our regulatory thinking.

This is a book for those studying, or about to study, law – as well as others with interests in the legal, political, and social impact of technology.

Roger Brownsword is Professor in Law at King's College London and at Bournemouth University, Honorary Professor at Sheffield University, and Visiting Professor at City University Hong Kong.

Law 3.0

Rules, Regulation, and Technology

Roger Brownsword

Routledge
Taylor & Francis Group
a GlassHouse Book

First published 2021
by Routledge
2 Park Square, Milton Park, Abingdon, Oxon OX14 4RN

and by Routledge
605 Third Avenue, New York, NY 10017

A GlassHouse Book

Routledge is an imprint of the Taylor & Francis Group, an informa business

British Library Cataloguing-in-Publication Data
A catalogue record for this book is available from the British Library

Library of Congress Cataloging-in-Publication Data
Names: Brownsword, Roger, author.
Title: Law 3.0 : rules, regulation and technology / Roger Brownsword.
Other titles: Law three point zero
Description: Abingdon, Oxon ; New York, NY : Routledge, 2020. |
    Includes bibliographical references and index.
Identifiers: LCCN 2020007916 | ISBN 9780367488635 (paperback) |
    ISBN 9781003053835 (ebook)
Subjects: LCSH: Technology and law. | Law—Methodology.
Classification: LCC K487.T4 . B76 2020 | DDC 344/.095—dc23
LC record available at https://lccn.loc.gov/2020007916

ISBN: 978-0-367-51640-6 (hbk)
ISBN: 978-0-367-48863-5 (pbk)
ISBN: 978-1-003-05383-5 (ebk)

Typeset in Times New Roman
by Apex CoVantage, LLC

Contents

Chapter 1

Introduction to Law 3.0

This book is an introduction to what I will call 'Law 3.0.' While Law 3.0 is a particular kind of conversation (and mindset), to be compared and contrasted with two other conversations (and mindsets), Law 1.0 and Law 2.0, it is also a shorthand for an extended field of legal interest – a field that features these three coexisting and interacting conversations (and mindsets).

To illustrate Law 3.0, recall the disruption at London Gatwick airport shortly before Christmas 2018, when an unauthorised drone was sighted in the vicinity of the airfield. As a precautionary measure, all flights were suspended, and for two days the airport was closed. With thousands of passengers stranded and inconvenienced, this was headline news, and if such bad news stories were to be avoided, it was clear that some action would need to be taken. But how should the law respond?

Representatives of the pilots' association told television interviewers that they had been saying for some time that drones presented a real danger to aircraft and that the exclusion zones around airfields (set by the current legal rules) needed to be extended. In due course, the government responded by announcing that the police would be given new powers to tackle illegal drone use and that the drone no-fly zone would be extended to three miles around airports. By so doing, the government recognised that the existing rules were not fit for purpose (not fit to protect the safety of aircraft, their pilots, crew, and passengers) and that changes needed to be made.

At the same time, however, others responded rather differently. Instead of focusing on whether the legal rules were fit for purpose, they focused on the possibility of finding a technological solution, ideally one that rendered it impossible in practice for a drone to be flown near an airport (or, failing that, a technology for disabling and safely bringing down unauthorised drones). In other words, rather than relying on rules to manage the risks associated with air travel, they sought to improve the design and safety specification of drones and/or airfields.

In these two responses, focusing on both rule changes and technological solutions, we see the essential features of a Law 3.0 conversation. Legal rules need to be updated and revised so that they are fit to serve their intended purposes or policies – whether these purposes and policies relate to health and safety, or climate change, or protecting consumers, or controlling crime, or whatever. Moreover, the apparatus and resources that support these rules – the supervisory and enforcement agencies – need to be sustained and upgraded so that the rules are fit for purpose, not only on paper but also in practice. Alongside the use of rules, however, we should also seek out possible technical solutions. When we say 'technical' or 'technological' solutions, this covers a broad range of measures that might supplement or supplant the rules. The measures might be 'architectural,' so that buildings and spaces are designed to reduce the opportunities for crime, or accident and injury, or the unnecessary use of energy, and the like. They might also be incorporated in the design of products or processes (simply by automating a process, humans might be removed from potentially dangerous situations), and, in principle, the technical measures might be incorporated in wearables or even in humans themselves.

We find a similar two-pronged approach – rules fit for purpose plus technological solutions – in the government's proposed strategy for dealing with the kind of online content that is clearly problematic. For example, what measures should be taken to tackle content that might compromise national security or encourage terrorist acts, or that threatens and endangers politicians, or that targets vulnerable parties, such as children, the elderly, the addicted, and so on? First, it is recognised that the legal rules need to be rendered fit for purpose in the digital age (notably by establishing a new statutory duty of care on Internet companies to take reasonable steps to keep their users safe and tackle illegal and harmful activity on their services), and second, it is proposed that effective technical measures should be explored, the aspiration being to make technology itself a part of the solution.

In this light, imagine a Law 3.0 response to the persistent problem of motorists who use their mobile phones while driving. On the one hand, the rules and the penalties need to be fit for purpose, and sufficient resources need to be committed to policing compliance with the rules. On the other hand, there are some technical options. One is to build hands-free mobile phone facilities into cars, but if motorists are distracted by using a phone, even a hands-free phone, this is not the complete answer. Possibly, there are ways of disabling phones while they are in vehicles, but if this means that passengers, even in an emergency, cannot use their phones, then again this is not the complete answer. Happily, with autonomous vehicles at an advanced stage of development, a technical solution is on the horizon: once automated vehicles relieve 'drivers' of their safety responsibilities, it seems that the problem will drop away – rules that penalise humans who use their mobile phones while driving will become redundant. Humans will simply be transported in vehicles, and the one-time problem of driving while phoning will no longer be an issue.

Of course, the conversation and mindset that is Law 3.0 did not come from nowhere: before Law 3.0, there was Law 2.0, and before Law 2.0, there was Law 1.0. However, it should not be thought that Law 3.0 meant the end of Law 2.0 or indeed the end of Law 1.0. Instead, new conversations emerged with new technologies but without wholly eclipsing earlier conversations. Law 3.0, as a distinctive conversation, coexists with Law 1.0 and Law 2.0, and Law 3.0, as the reconfigured field of legal interest, comprises these three coexistent conversations. What, then, do we understand by Law 1.0, and how did Law 1.0 evolve to provoke Law 2.0, which in time evolved to provoke Law 3.0?

Characteristically, the form of reasoning in Law 1.0 is the application of rules, standards, and general principles to particular fact situations. Some of the rules, standards, and principles will be fairly elastic – any legal provision that includes the word 'reasonable' is geared for flexibility – which means that Law 1.0 does have some capacity to make adjustments for novel situations and to make exceptions in response to those notorious hard cases where the application of the relevant rule simply does not seem fair in the circumstances. Indeed, in a recent statement aimed at clarifying the legal status of cryptoassets and smart contracts (giving guidance on whether, in principle, they can be treated respectively as 'property' and 'contracts'), the UK Jurisdiction Taskforce (associated with the LawTech Delivery Panel) (2019: para 3) has claimed that the 'great advantage of the English common law system is its inherent flexibility'; that 'judges are able to apply and adapt by analogy existing principles to new situations as they arise'; and that '[t]ime and again over the years the common law has accommodated technological and business innovations, including many which, although now commonplace, were at the time no less novel and disruptive than those with which we are now concerned.' Nevertheless, the greater the industrialisation of a society, the more that technologies are employed, the greater the strain on Law 1.0, because it is simply not geared to respond to the range of risks that now present themselves.

With Law 1.0 under stress, a natural way of expressing discontent with the prevailing rules is to say that they are not fit for purpose – specifically, that they are not fit for the purposes or policies that are being adopted for the further and better technological performance of society. At this point, Law 2.0 crystallises. Here, the form of reasoning is policy-directed and instrumental. The conversation is about what works. It is not a matter of recycling a body of traditional rules, standards, and principles; rather, it is a matter of articulating new rules and regulatory frameworks that directly serve the purposes which governments now adopt. If legal rules were like clothes, Law 2.0 would signal a whole new wardrobe.

In Law 2.0, the centre of gravity of law shifts from the courts and historic codes to the political arena where governments operate through the executive and the legislative assemblies. Accordingly, where new technologies, such as the aforementioned cryptoassets and smart contracts that rely on blockchain, need to be governed, in some legal systems the response will be a Law 2.0 conversation, but in others the conversation will start in Law 1.0 mode before moving to Law 2.0. As Sir Geffrey Vos (2019: para 5) put it when introducing the Jurisdiction Taskforce's statement, the approach 'has been to start from basic legal principles and work forward to regulation. . . . There is no point in introducing regulations until you properly understand the legal status of the asset class that you are regulating.'

The evolution from Law 2.0 to Law 3.0 builds on the instrumentalism of the former. However, it depends crucially on the perception that new technologies now present themselves as regulatory tools to be deployed alongside rules. To some extent, the idea that architecture and design can be used to defend person and property is as old as the pyramids and locks. However, the regulatory mindset in Law 3.0 is characterised by a *sustained* focus on the potential use of a range of technological instruments the density, sophistication, and variety of which distinguish our circumstances, quantitatively and qualitatively, from those of both pre-industrial and early industrial societies. Whether or not this amounts to a difference of kind or degree scarcely seems important; we live in different times, with a significantly different regulatory mindset and repertoire of technological instruments.

Consider one further example. In August 2019, at the start of the new premiership soccer season, there was much discussion about the use of VAR (video assistant referee) technology to assist on-field referees. In particular, the spotlight was on the use of VAR and the offside and handball rules. Most commentators thought that these rules were the real problem (that they were not fit for purpose) but it was agreed that VAR was exacerbating matters (by over-turning on-pitch decisions and 'goals' that had been celebrated in the stadium). In the interest of accurate application of the rules of the game – and people still debate whether the ball did fully cross the goal line for one of England's goals in the 1966 World Cup Final – VAR has a role to play, but it comes at a price. Thus, in one of his columns in *The Times*, Matthew Syed suggests that

> VAR was a car crash waiting to happen, a clear and obvious error, a technological innovation so obviously against the grain of the game that it is a wonder it was so widely touted. The only question now is: when are they going to bin it?
>
> (Syed, 2019)

To which, the answer is that we will bin VAR if we decide that we are more distressed by frustrated or delayed goal celebrations than by inaccurate refereeing decisions.

Law, it is often said, is rather like a game; in both law and games, there are rules, and in some games there are officials, and so on. However, we now see that there is more to this similarity than the fact that we are dealing with rule-governed activities. Nowadays, we see that Law 3.0 has its analogue in Soccer 3.0; in both law and soccer, the conversation is about both the fitness of the rules and the possibility of technical fixes. Moreover, where smart machines, such as Watson and AlphaGo, beat the best human game players, we see another aspect of Law 3.0, namely, the possibility of machines taking over various activities and functions previously performed by humans. That said, law is not entirely comparable to a game. Typically, we think that convicting an innocent person of a criminal offence is not to be compared with mistakenly disallowing a goal because a player was incorrectly judged to be offside.

In the year 2020, the idea persists that training a person 'to think like a lawyer' is training them to reason as in Law 1.0. Of course, Law 1.0 is what solicitors do when they advise a client on the legal position, it is what barristers do when they draft an opinion for a client, and it is the form in which common law judges justify their decisions. As we have already noted, it is also the way in which the Jurisdiction Taskforce framed its inquiry and statement. Nevertheless, there are some awkward questions when Law 1.0 meets Law 2.0 – for example, questions about the relationship between principle (Law 1.0) and policy (Law 2.0) and about the line between law and politics. Similarly, when Law 1.0 meets Law 3.0, as it increasingly will, there is some awkwardness, but it is exacerbated by questions about the coexistence of legal principles with expanding technological possibilities. However, Law 3.0 is where we are; Law 3.0 is the conversation to be in; and Law 1.0, by largely ignoring technology as a solution to regulatory problems, is in danger of consigning itself to becoming little more than a sideshow.

Stated shortly, this book, as an introduction to Law 3.0, is about the disruption of law and legal reasoning by new technologies as a result of which, I suggest, there is a need to reimagine and then to reinvent law. It is about the disruptive impact of new technologies on the traditional content of legal rules, about the disruption and displacement of the way that those associated with the legal and regulatory enterprise reason, and about the increasing availability of technological instruments to support, or even supplant, legal rules.

The argument is that, in the wake of this disruption, there is a need to reimagine the field (the regulatory environment) of which legal rules are a part. Instead of thinking exclusively in terms of a certain set of rules and norms (representing 'the law'), it is suggested that we should think of a set of tools that can be employed for regulatory purposes. While some of these tools (such as legal rules) are normative, others (employing, for example, the design of products or processes) are non-normative. While normative instruments always speak to what 'ought' to be done, non-normative instruments – at any rate, at the hard end of the spectrum – speak only to what 'can' and 'cannot' be done.

Having got our heads around the idea of Law 3.0, how are we to live with it? Who is to be invited to the Law 3.0 conversation, and what kind of conversation does it need to be? My argument is that, if law is to be reinvented, the renewal should be anchored to a new foundational understanding of regulatory responsibilities (and, concomitantly, a 'triple licence' for the use of technological instruments) on which we can then draw in order to shape our articulation of the Rule of Law, in order to revitalise 'coherentist' Law 1.0 thinking, and in order to refashion legal and regulatory institutions both locally and internationally.

My conclusion is not that, with law so reinvented, all will go well. In a world of dynamic technological change, maintaining the conditions that are essential for human social existence will always be a challenge, and discharging our regulatory responsibilities will inevitably be a work in progress. Nevertheless, I suggest that the chances of things going well are somewhat better if we do reimagine and then reinvent law rather than take no steps in this direction.

Finally, let me close these introductory remarks by saying two things about the way in which this book is written and presented. The first thing is that, in writing this book, I am departing from the usual scholarly conventions by eliminating footnotes, by keeping my references to an absolute minimum, and by trying to write short chapters. I want the text to speak for itself so that it can be read quickly, understood, and retained. I should also say that, although the law is no stranger to making use of fictions, readers should be prepared for some chapters (such as the next one) that indulge in pure fiction. For readers who want to follow up with a more traditional academic presentation of the main ideas in this book, they can find this in my earlier monograph, *Law, Technology and Society* (Brownsword, 2019a). The same is true of what I say about legal education in Chapter 25 of the present book (see further Brownsword, 2019b). The second thing is that I would not want my presentation to mislead readers into supposing that the field of legal scholarship that we now recognise as that of 'law, regulation, and technology' (Brownsword and Yeung, 2008; Brownsword, Scotford, and Yeung, 2017; Guihot, 2019) is short of contributors and light on literature. To the contrary, there are many scholars worldwide contributing to both the breadth and the depth of what is a burgeoning literature. Accordingly, right at the end of the book, I make some short comments on the way in which the literature has grown over the last 30 years – writers first responding to developments in biotechnology, computing, and information and communications technology, then to developments in nanotechnology and neurotechnology, and more recently to additive manufacturing, blockchain, and artificial intelligence and machine learning – after which I offer some suggestions for further reading.

Chapter 2

BookWorld

A short story about disruption

Imagine a (fictitious) local independent bookshop. We can call it BookWorld. The bookshop has a special place in the life of the local community. BookWorld is more than just a bookshop. On one occasion, when BookWorld was moving from its old store to new premises nearby, the community famously formed a human chain to remove the stock of books from one site to the other. At BookWorld, books are shelved in accordance with a classificatory scheme that has served it, and its community of book-lovers, well. The scheme starts with fiction and non-fiction, but then it employs various subclasses. Occasionally, there will need to be some discussion about the right place to shelve a particular book, but this is pretty exceptional. In general, staff and customers alike know where to find the titles in which they are interested.

However, with the explosion of books about new technologies – starting with books about biotechnologies and cybertechnologies, but now including books about neurotechnologies and nanotechnologies, about AI and machine learning, about virtual reality and autonomous vehicles, and about blockchain and 3D printing, and so on – the staff at BookWorld have found it difficult to know how to respond. Should these titles be shelved under 'Popular Science,' or 'Medicine,' or 'Health,' or 'Economics,' or 'Law,' or 'Ethics,' or 'Smart Thinking,' or even 'Science Fiction'? While staff can advise customers whether or not they have a particular title in stock, they cannot always be sure where a particular book is shelved; for, quite literally, the titles on emerging technologies are 'all over the shop.' Moreover, staff realise that it is not possible to find a home for these burgeoning titles on technologies and their applications without stretching and distorting the classificatory indicators, or without creating ad hoc classes. There is a reluctance to revise the traditional classificatory scheme, but as the technology titles attest, the scheme is no longer fit for purpose.

Eventually, BookWorld yields to the inevitable: one room in the store is now dedicated to 'emerging technologies' within which titles are shelved in accordance with a brand-new classificatory scheme. However, this is not the end of the problems for BookWorld. The bookshop is experiencing a further, and more radical, disruption. Once upon a time, a book was a book, and a bookshop was a bookshop; but now the formats for books, and for selling books, are varied. Books are now supplied in digital formats, and 'virtual' bookstores have only an online presence.

As the owners of BookWorld realise all too well, many of the technologies about which the books are written are themselves exerting a disruptive influence on retailing generally and on bookstores in particular. In this context, sales at Book-World are down and, even though their customers are loyal, that loyalty is challenged when online sellers can deliver titles so quickly and at such competitive prices. Putting it bluntly, the owners of BookWorld realise that they need to rethink the business. They need to review their objectives and reimagine how the bookstore might serve their purposes. Should BookWorld be reimagined as an online bookstore or as an online bookstore with a brick-and-mortar annex; or as a brick-and-mortar bookshop but with an online option; or should it follow the lead of the (also fictitious) Fully Automated Bookstore (FAB); or should it continue to trade exclusively as a modernised but still relatively traditional brick-and-mortar store?

Historically, BookWorld has invested heavily in building long-term relationships with its customers. The owners take pride in acting on the basis that there is more to their business than the selling of books. Although online businesses seem to be the future, closing the store and moving the business online simply is not an option. BookWorld is part of the fabric of the community; it is part of the infrastructure for the social life of many people; and the owners are all too aware that, in too many town centres, there are too many empty retail units. Somehow, Book-World needs to reinvent itself so that it incorporates the best of the traditional business model and the benefits of a raft of new technologies. What to do? What the owners are after seems to be a judicious and socially acceptable combination of the traditional and the technologically enabled bookstore. The owners do not know what particular combination that is, but in an effort to reinvent BookWorld in a way that is acceptable to all stakeholders, the owners decide that they should consult their customers and the local community.

Having taken the story this far, I can invite readers to write their own ending. The disruption of BookWorld leads to a necessary exercise in reimagination, but there is no guarantee that the owners of the bookstore will be successful in reinventing the shop. Disruption, reimagination and reinvention is not a dystopia, but neither is it a process that necessarily ends well. The most that we can say is that the owners seem to be proceeding in an enlightened way and that we hope that this story ends well for BookWorld and its customers.

But, irrespective of the fate of BookWorld, what does this tell us about the law? Across the street from BookWorld, a family law firm is experiencing a similar kind of disruption. These days, the imposing leather-bound statutes and law reports that line the bookshelves in the office are more for decoration than reference; the burgeoning range of legal services offered by online providers has significantly reduced the number of people who walk in to the office seeking legal advice and assistance, and even established clients are being drawn away by law firms that have invested heavily in the automation of document drafting, disclosure, due diligence, and so on. Whether we are looking at the sale of books or the provision of legal services, traditional practices have been disrupted by new technologies.

Though the development of LawTech is important, the disruption to traditional ways of providing legal advice and assistance is not the principal lesson to be taken from the story of BookWorld. That lesson instead is about the disruption to the traditional Law 1.0 mindset, a mindset evinced by a reluctance to revise classificatory schemes even when they are obviously unfit for purpose. This disruption leads, in Law 2.0 and Law 3.0, not only to new classificatory schemes but to a radically different appreciation of how technological tools might contribute to the operationalisation of bookshops (and the law). In this way, BookWorld tells us something highly significant about the evolution of legal thinking and about how technology increasingly affects the manner in which we perform what the American jurist Karl Llewellyn called the 'law-jobs' (Llewellyn, 1940). In particular, it flags up the increasing relevance of technology for, in Llewellyn's terms, the channelling (or regulating) of human conduct.

Finally, just as with BookWorld, the process of disruption, reimagination and reinvention of law is not necessarily destined to be dystopian – but neither is it guaranteed to end well. However, if we go about it in the right way, we can certainly hope, and perhaps even expect, that the reinvention of law, Law 3.0, will be a success story.

Part one

The technological disruption of law

Chapter 3

Law 1.0

Easy cases, difficult cases, and hard cases

In the introductory chapter, we said that the conversation in Law 1.0 is about applying the general principles of the law (and its more particular rules) to specified fact situations. On such and such facts, law students are asked, what would be the legal position? Sometimes, the applications of the legal rules and principles are straightforward; sometimes they are more difficult. Sometimes it is a technological innovation (anything from a bicycle to an autonomous vehicle) that creates the difficulty, but often it is not technology that is the source of the difficulty. In other words, Law 1.0 has its challenges quite apart from technological innovation. To bring legal rules and principles to bear on novel fact situations, whether or not they involve technology, requires a degree of imagination. However, as we will explain in the next chapter, it is when our imagination is challenged by new technologies that Law 1.0 is likely to be disrupted.

As students of the law soon learn, it was in the middle years of the nineteenth century that the common law courts laid down the general principles to be applied when making awards of damages for breach of contract. The master principle was that the object of damages was compensatory with a view to putting the innocent party in the same position as they would have been in had the contract been performed (i.e., had there been no breach). At the same time, the courts also set out the principles to be applied where the claim was for so-called consequential losses flowing from the breach of contract. Stated simply, defendants would be liable to compensate for such losses provided that they arose in the ordinary course of things (such that anyone would anticipate them) or, where the consequential losses were 'extraordinary,' the defendant (in the light of their knowledge as a contracting party) could anticipate them.

In Law 1.0 conversations about breach of contract, these are key principles to be applied. However, while there are many 'easy cases' where applying these principles to the particular facts is straightforward, there are also 'difficult' cases where applying the law is less straightforward as well as occasional 'hard cases' where the principle is settled and its application straightforward but the result of the application gives us pause (typically because it seems unreasonable or unjust in the particular case).

If, for example, a seller in breach of contract fails to deliver the goods as agreed, the buyer is expected to buy from another seller paying whatever the market price is at the time. Where the market price is higher than the original contract price, the compensatory principle indicates that the damages awarded should cover the difference between the market price and the original contract price. That way, the buyer is placed in the same position (at least, financially) as they would have been in had the seller performed the original contract. So far, so easy: the applicable principle is settled, the application is straightforward and the result is unproblematic.

However, there might be problems in relation to any one of the three elements of an easy case. In other words, the applicable principle might not be agreed and settled; the application of an agreed and settled principle might not be straightforward; and the result of applying the principle might not be acceptable.

Where the applicable principle is not agreed and settled, where there is more than one candidate rule or principle, we have a serious problem in relation to the 'coherence' of the law (about which we will say much more in later chapters). Although, as we have said, the general principle relating to the recovery of consequential losses was settled in the nineteenth century, it has become unsettled in the twenty-first century. Today, it is not clear whether the governing principle hinges on what the defendant contract breaker might reasonably have anticipated as a result of the breach or whether it turns on the defendant having 'assumed responsibility' for the losses in question.

Where the applicable principle is agreed and settled, there might still be more than one way of interpreting and applying it. Notoriously, this is so with the general compensatory principle. For example, if a contractor installs a kitchen, but in breach of contract supplies a cupboard unit that is fractionally out from the contract specification, how is the client to be put in the same position as if the contract had been performed? In the context of a commercial contract, it might be sufficient to award the difference in value between the kitchen as specified and the kitchen as constructed, but in a consumer contract, where the consumer has a personal interest in the specification, a more reasonable award might be the cost of correcting the breach. However, where curing the breach involves demolishing some of the kitchen already built, this could be very expensive (not to mention, as some would see it, 'wasteful'), and there might then be a huge gap between the difference in value and the cost of cure. This is now a difficult case and judges might reasonably disagree about how to apply the compensatory principle.

So, for example, in the leading English case of *Ruxley Electronics v Forsyth* (1996), where a swimming pool was built nine inches shallower than specified by the contract, it was held that there was no difference in value between the pool as constructed (or the private property in which it was built) and the pool as it should have been constructed, but that it would cost in excess of £20,000 to rectify the shortfall. The pool was also found to be perfectly safe for swimming and diving, but it was not in accordance with the client's specification. The trial judge awarded the client £2,500 damages for breach of contract, for so-called loss of amenity (neither the difference in value nor the cost of cure); on appeal, the majority of the Court of Appeal awarded the client the full cost of cure. But then, on final appeal, the House of Lords restored the trial judge's award. Everyone agreed that the client should be put in the same position as if the swimming pool had been constructed in accordance with the contract, but there were two recognised ways of applying that principle (awarding damages measured either by reference to the difference in value or by reference to the cost of cure), and in the event, the House of Lords did neither, coming up (like the trial judge) with a third way of applying the principle (by reference to the loss of amenity).

The third possibility is that the result of a straightforward application of the agreed principle is not acceptable. For instance, if a contract breaker makes a financial gain by breaking the contract but there is no financial loss to the innocent party, should the former be accountable to the latter for the gain made? In many business contexts, the fact that one party takes advantage of a better offer might not cause us too much concern, but where the breach is egregious and clearly one that the innocent party would not have waived, we have a hard case. Such cases are highly problematic because a sympathetic response to the merits of the case might then create doctrinal confusion and incoherence.

While the classic hard case is about a result that is unreasonable or unjust in the particular circumstances, such cases might be symptomatic of a more general problem with the law – and sometimes it will be because a new technology is harmful in ways for which the general principles are not adequately geared to compensate. Consider, for example, the compensatory claims made by parties whose reproductive plans have been frustrated by negligent acts, negligent advice, negligent omissions, and so on of those medical professionals who deal with new reproductive techniques and tests. In his book, *Birth Rights and Wrongs,* Dov Fox sketches the landscape in the following way (Fox, 2019: 165–166):

> Different kinds of reproductive wrongs call for different kinds of rights. In some cases, procreation is *deprived* – as when a lab technician drops the tray of embryos that are an infertile couple's last chance to have biological children, or when a doctor leads an eagerly expecting pregnant woman to abort by misinforming her that her healthy fetus would be born with a fatal disease. In other cases, procreation is *imposed* – as when a pharmacist fills a woman's birth control prescription with prenatal vitamins, or when a surgeon botches the sterilization that parents of five had sought because they were already struggling to make ends meet. Procreation is *confounded* when an IVF clinic fertilizes a patient's eggs with sperm from a stranger instead of her spouse, or when a sperm bank neglects to inform prospective parents that the anonymous donor it called 'perfect' had actually dropped out of college, been convicted of burglary, and diagnosed with schizophrenia.

Although some courts have responded positively and imaginatively to such cases, Fox is critical of the widespread failure of US courts to do so. Where the negligence at issue involves the loss or destruction or mis-transfer of embryos, courts tend to decline to compensate because the resulting 'harm' or 'loss' does not fit with the usual understanding of physical damage or damage to 'property,' and where the negligence involves failing to prevent a pregnancy or the birth of a child with a particular inherited condition, the courts tend to hold that the birth of a child (even an unplanned child) is a cause for celebration rather than compensation.

Where the courts cannot, or do not, use their imagination to rework tort principles, or traditional ideas of what can count as property, or to invoke notions of privacy or human dignity (compare Brownsword, 2003), to apply to such cases, we have reached the limits of Law 1.0 and a Law 2.0 conversation and response is invited.

Chapter 4

Law 1.0 disrupted

The story of the technological disruption of Law has two stages: first, new technologies disrupt Law 1.0 and encourage the emergence of Law 2.0, and second, Law 2.0 is disrupted by the availability of technologies as regulatory tools, leading to Law 3.0. In this chapter, the focus is on the first of these disruptions.

The first disruption causes us to question the adequacy of existing rules of law (we begin to wonder, as we would now put it, whether these rules are fit for purpose). Such disruption can highlight more than one form of unfitness.

One form of unfitness is seen where the substance of prevailing legal rules no longer is appropriate relative to the desired regulatory purposes; the rules at issue need to be changed. This was the case, for example, at the time of the drone-related shutdown at Gatwick Airport: it was recognised that the rules that set exclusion areas for drones around airports were not adequate and that they needed to be changed. Similarly, we might respond to the restricted remedies available for negligence where procreation is deprived, imposed, or confounded (as discussed in the previous chapter) by arguing that the rules need to be corrected.

Another form of unfitness is seen where the prevailing legal rules simply make no provision for the technology or its application; in other words, the deficiency takes the form of a gap or an omission. To rectify the problem, there needs to be a bespoke regulatory response. For example, if there had been no rules setting exclusion areas for drones, then after the Gatwick Airport incident there would have been calls for appropriate rules to be introduced. Similarly, with reference to the burgeoning range of reproductive technologies, there might be calls for more than 'tweaking' the common law principles; it might be argued that a comprehensive regulatory framework for, among other things, the practice of IVF needs to be adopted.

Yet a further form of unfitness is where the rules no longer map onto, or connect with, the technology and its applications. For example, if the rules assume that drones (or, likewise, vehicles or vessels) will be operated by identifiable persons, they will not recognise drones (or vehicles or vessels) that are fully autonomous. There is a 'disconnect' between the rules and the actuality.

Through the last two centuries, new technologies and their applications have systematically disrupted some of the cornerstone principles and rules of Law 1.0, highlighting the lack of fitness of traditional standards for societies that are undergoing rapid industrialisation. As Geneviève Viney and Anne Guégan-Lécuyer put it, a tort regime 'which seemed entirely normal in an agrarian, small-scale society, revealed itself rather quickly at the end of the nineteenth century to be unsuitable' (2010: 50). On the one hand, a tort regime centred on individual fault and personal responsibility (putting the onus on claimants who were injured in accidents at work, on the railways, and so on, to show that there had been a failure to take reasonable care) was too demanding; and irrespective of fault, in the absence of modern insurance schemes, individual defendants were unlikely to be in a position to meet the compensatory needs of victims. On the other hand, there was a risk that fault-based regimes would overexpose nascent enterprises and discourage innovation. In short, from a regulatory perspective, the rules of Law 1.0 were under-protective in relation to both injured persons and innovative enterprises, and the challenge set by Law 2.0 was to create a regulatory environment that responded to these deficiencies.

Meanwhile, in the nineteenth century, there was a dramatic shift from the Law 1.0 idea that a necessary element of a crime is that the offender has a guilty mind and the relevant intent (so-called, *mens rea*). Breaking with Law 1.0, in both the United Kingdom and the United States, there was a steady growth in criminal offences that were punishable without criminal intent. The world was changing, and new technologies were driving the changes. As Francis Sayre (1933: 68–69) remarked, the

> invention and extensive use of high-powered automobiles require new forms of traffic regulation; . . . the growth of modern factories requires new forms of labor regulation; the development of modern building construction and the growth of skyscrapers require new forms of building regulation.

So it was that the courts accepted that, so far as 'public welfare' offences were concerned, it was acceptable to dispense with proof of intent or negligence. If the food sold was adulterated, if vehicles did not have lights that worked, if employees polluted waterways, and so on, sellers and employers were simply held to account. For the most part, the penalty was a fine, which might be regarded as no more than a tax on business; it relieved the prosecutors of having to invest time and resources in proving intent or negligence. As Sayre reads the development, it reflected 'the trend of the day away from nineteenth century individualism towards a new sense of the importance of collective interests' (67).

In the case of contract law, the key moments of disruption start with a shift from a 'subjective' consensual model of agreement to an 'objective' approach. The idea that contractors have to be subjectively ad idem, actually to have agreed on the terms and conditions of the transaction, hampered enterprises that needed to limit their liabilities associated with new transportation technologies. In the jurisprudence, this shift is epitomised by Mellish LJ's direction to the jury in *Parker v South Eastern Railway Co* (1877), where the legal test is said to be not so much whether a customer actually was aware of the terms and had agreed to them but whether the railway company had given reasonable notice. About a hundred years later, we come to a second moment of disruption when, with the development of a mass consumer market for new technological products (cars, televisions, kitchen appliances, and so on), it was necessary to make a fundamental correction to the traditional values of 'freedom of contract' and 'sanctity of contract' in order to protect consumers against the small print of suppliers' standard terms and conditions. Today, new technologies for transactions (such as online environments for commerce and blockchain) continue to disrupt the law, but these are largely questions for Law 2.0 and Law 3.0 rather than Law 1.0.

What we see across these developments is a pattern of disruption to legal doctrines that were organically expressed in smaller-scale non-industrialised communities – communities where horses, not machines, did the heavy work. Here, the legal rules presuppose very straightforward ideas about holding to account (moreover, holding *personally* to account) those who engage intentionally in injurious or dishonest acts, about expecting others to act with reasonable care, and about holding others to their word. Once new technologies disrupt these ideas, we see the move to strict or absolute criminal liability without proof of intent, to tortious liability without proof of fault, to vicarious liability (particularly holding employers liable for the careless acts of their employees), and to contractual liability (or limitation of liability) without proof of actual intent, agreement, or consent. Moreover, these developments signal a doctrinal bifurcation, with some parts of criminal law, tort law, and contract law resting on traditional principles (and representing, so to speak, 'real' crime, tort, and contract) while others deviate from these principles as necessary adjustments or corrections are made.

Alongside disruption to the existing rules, we also have cases where the development or application of a new technology exposes gaps or omissions in the law. Historically, we can identify many instances in which new technologies have opened up places (for example, airspace, the seabed, and the polar regions) to exploitation by humans (for both military and commercial purposes) as a result of which there has been a pressure to adopt a governing set of ground rules. More recently, developments in both biotechnologies and cybertechnologies have exerted a similar pressure. For example, a legal framework has to be created to lay down the ground rules for the provision of, and access to, the latest technologies for assisted reproduction; new offences have to be created to deal with a range of matters from human reproductive cloning to cybercrime; the development of computers also necessitates setting out a legal framework for the processing of personal data; and there needs to be some gap-filling and stretching of IP law to cover such matters as databases, software, and integrated circuits as well as modern biotechnologies whose working cannot be demonstrated to patent examiners by taking a machine model into the Patent Office. What is distinctive about this kind of disruption is not so much that there are additions to the legal rule book but that these responses are typically bespoke, tailored, and in a legislative form, and critically, the regulatory mindset that directs these responses is quite different to traditional coherentist patterns of thought.

Finally, rules can be disrupted because they simply do not map onto, or con-nect with, the technology or its applications. Drones that are fully autonomous might be a case in point, but the examples that most obviously spring to mind are autonomous vehicles and autonomous vessels. In both these cases, the road traffic and maritime rules that assume that an in-vehicle or an on-vessel person will be in control (the driver or the master) no longer fit the facts. Similarly, there might be some questions about how existing classifications and concepts map onto the operation of still human-driven Uber cars. For example, given the geo-locating features of the Uber app, there might be a question about whether these cars are in effect 'plying for hire' in breach of the restrictions on private hire cars (see *Reading Borough Council v Mudassar Ali*, 2019, where the court actually answered the question in the negative). Coherentists might make heroic attempts to apply the existing rules, to make the existing rules fit, but sooner or later the disruption will have to be addressed head-on and a bespoke regulatory response made.

These disruptions to the rules may take different forms. However, the key point is not so much the various ways in which the application of general principles and rules is challenged as the nature of the response. When Law 1.0 is disrupted, it is not just the rules of law that are disturbed; the disruption is also to the coherentist mindset of Law 1.0. By contrast, in Law 2.0, we have a different conversation, a different mentality, a different process, and different rule products.

Chapter 5

Law 2.0 and technology as a problem

The disruption of Law 1.0 prompts the development of a regulatory-instrumentalist mindset. The Law 2.0 conversation is not about the internal coherence or the application of general legal principles but about whether the rules are fit for purpose in responding to emerging technologies. On the one hand, the rules will be unfit if they involve over-regulation, stifling the development and application of beneficial new technologies, but on the other hand, the rules will be unfit if they involve under-regulation, exposing persons to unacceptable risks (whether of a physical, psychological, financial, or other nature) or compromise values that are important in the community.

For regulators who aspire to get it right, a Law 2.0 mindset might pose the right questions, but with rapidly developing and contested technologies, the answers can be elusive. Technology, in short, is a problematic target for regulators.

Here we can speak to three particular respects in which technology is a challenge for regulators who think and operate in a Law 2.0 fashion. These three challenges relate to the legitimacy of the regulatory position, the connectedness of the regulation, and the effectiveness of the regulatory action (or inaction).

Regulatory legitimacy

While the regulatory environment for each technology will reflect a mix of local politics, preferences, and priorities, we can identify three generic desiderata – or, at any rate, these are desiderata for communities in which citizens expect to enjoy the benefits of innovation but also expect technologies to be safe and to be applied in ways that respect fundamental values. Here, regulators will face a triple demand:

* to support rather than stifle beneficial innovation
* to provide for an acceptable management of risks to human health and safety and the environment
* to respect fundamental community values (such as privacy and confidentiality, freedom of expression, liberty, justice, human rights, and human dignity)

The challenges of these three demands reside both in the tensions *between* them and the tensions hidden *within* them.

Regulators will find that, while the innovation lobby will argue for light-touch regulation, for strong intellectual property rights, for tax breaks, for subsidies, and so on, other parties will argue that (a) there need to be proper *ex ante* risk assessments and precautions in place and (b) adequate regulatory oversight might be needed to protect fundamental values. This, it will be said, demands a 'proportionate' response by regulators, weighing the burden on innovators (and, possibly, the delayed public enjoyment of benefits) against community concerns for safety and respect for values. This, of course, simply restates the challenge without resolving it.

Tensions *between* the demands aside, each of the demands hinges on a deeply contested concept. In the case of the first demand, we should ask: what kind of innovation is 'beneficial'? Beneficial to whom, beneficial in meeting whose needs, beneficial relative to which human interests? Beneficial when – at once, within the next five years, or at some unspecified time in the future?

With regard to the second demand, we should ask: what is an 'acceptable' risk and to whom is the burden of risk 'acceptable'? Notoriously, the view of professional risk-assessors differs from the lay view in characterising a technology as 'low risk' (inviting a leap to 'safe') as long as the likelihood of the harm is low, even though anyone would see the harm in question as extremely serious (for example, a commercial air crash is very rare but typically deadly). Moreover, how is the risk distributed? Who benefits and who bears the risk?

Last, but not least, which values (and which particular conception of a value) are to be treated as guiding? Which value system do we support – one based on rights, one based on duties, or one geared towards maximising utility? If we base ourselves on rights, then which rights (negative only or negative and positive, libertarian or liberal-welfare, and so on)? If on duties, then which duties (for example, Kantian or communitarian)? If on utility, then which variant (act or rule, ideal or non-ideal, and so on) do we adopt? If, instead, we are guided directly by values such as privacy or human dignity, liberty or justice, equality or solidarity, then which of the many conceptions of these values are to be taken as the reference standard?

Regulatory connection

One of the distinctive challenges presented to regulators by rapidly developing modern technologies is, quite simply, the pace of their development. How do regulators get connected to these technologies, and how do they stay connected? As John Perry Barlow (1994) famously remarked:

> Law adapts by continuous increments and at a pace second only to geology in its stateliness. Technology advances in . . . lunging jerks, like the punctuation of biological evolution grotesquely accelerated. Real world conditions will continue to change at a blinding pace, and the law will get further behind, more profoundly confused. This mismatch is permanent.

Whether one looks at the regulation of information technology or the regulation of biotechnology – or, for that matter, at the regulation of nanotechnology or the technologies associated with the new brain sciences, let alone blockchain, AI, and machine learning – there seems to be ample support for Barlow's thesis. Indeed, it is arguable that the pace of technological development, already too fast for the law, is accelerating. While this is not an easy matter to measure, there are at least two respects in which modern information technology, in addition to being significant in its own right, plays a key enabling role relative to other technologies – facilitating basic research in biotechnology (spectacularly so in the case of sequencing the human genome) as well as the commercial exploitation of the products of other technologies.

Technology is capable of leaving the law behind at any phase of the regulatory cycle: namely, before regulators have anything resembling an agreed position, before the terms of the regulation are finalised, and once the regulatory scheme is in place. For example, a new technology might emerge very quickly, catching regulators (at any rate, national legislators) cold; or it might be that a controversial new technology develops and circulates long before regulators are able to agree upon the terms of their regulatory intervention. While regulators are getting up to speed, or pondering their options and settling their differences, the technology moves ahead, operating in what for the time being at least amounts to, if not a regulatory void, at least a space in need of regulatory attention. As Michèle Finck (2019: 64) (thinking about the Internet and, potentially, distributed blockchain technologies) rightly remarks: 'When systems with regulation-defiant features are adopted on a large scale, social norms will shift to reject regulatory intervention. In such a setting regulation not only becomes hard from a technical perspective; it also becomes politically unattainable.'

Moreover, even (or especially) when regulatory frameworks have been put in place, they enjoy no immunity against technological change. For example, the UK Human Fertilisation and Embryology Act 1990 was overtaken by developments in embryology (in particular, the ability to carry out genetic engineering in eggs which are then stimulated without fertilisation rather than in embryos) as well as by the unanticipated use of new embryo-screening procedures to identify embryos that would be tissue compatible with a born child needing a bone marrow transplant (the so-called 'saviour sibling' cases); and, notoriously, data protection laws are soon outpaced by both technological development and the purposes for which personal data are collected and processed (see, e.g., Swire and Litan, 1998).

Regulatory effectiveness

A good deal of research effort has been expended in tracking the impact of particular legal interventions. Some interventions work reasonably well, but many do not – many are relatively ineffective or have unintended negative effects (see Gash, 2016). Moreover, we also know that the cross-boundary effects of the online provision of goods and services have compounded the challenges faced by regulators. If we synthesise this body of knowledge, what do we understand about the conditions for regulatory effectiveness?

First, we appreciate that the problems might lie with the regulators themselves. For example, where regulators are corrupt (whether in the way that they set the standards, or in their monitoring of compliance, or in their responses to non-compliance), where they are 'captured' by regulatees, or where they are operating with inadequate resources, the effectiveness of the intervention will be compromised.

Second, it might be regulatees who are the problem. Generally, it seems that regulators do better when they act with the backing of regulatees (with a consensus rather than without it). The lesson of the well-known Chicago study, for example, is that compliance or non-compliance hinges not only on self-interested instrumental calculation but also (and significantly) on the normative judgements that regulatees make about the morality of the regulatory standard, about the legitimacy of the authority claimed by regulators, and about the fairness of regulatory processes (Tyler, 2006). However, regulatee resistance can be traced to more than one kind of perspective. Business people (from producers and retailers through to banking and financial service providers) may respond to regulation as rational economic actors, viewing legal sanctions as a tax on certain kinds of conduct; professional people (such as lawyers, accountants, and doctors) tend to favour and follow their own codes of conduct; the police are stubbornly guided by their own 'cop culture'; consumers can resist by declining to buy; and, occasionally, resistance to the law is required as a matter of conscience – witness, for example, the peace tax protesters, physicians who ignore what they see as unconscionable legal restrictions, members of religious groups who defy a legally supported dress code, and the like.

In all these cases, the critical point is that regulation does not act on an inert body of regulatees: regulatees will respond to regulation – sometimes by complying with it, sometimes by ignoring it, sometimes by resisting or repositioning themselves, sometimes by relocating, and so on. Sometimes those who oppose the regulation will seek to overturn it by lawful means, sometimes by unlawful means; sometimes the response will be strategic and organised, and at other times it will be chaotic and spontaneous. But, regulatees have minds and interests of their own; they will respond in their own way, and the nature of the response will be an important determinant of the effectiveness of the regulation.

Thirdly, the problem might be various kinds of external distortion or interference with the regulatory signals. Some kinds of third-party interference are well-known – for example, regulatory arbitrage (which is a feature of company law and tax law) is nothing new. However, even where regulatory arbitrage is not being actively pursued, the effectiveness of local regulatory interventions can be reduced as regulatees take up more attractive options that are available elsewhere.

Although externalities of this kind continue to play their part in determining the fate of a regulatory intervention, it is the emergence of the Internet that has most dramatically highlighted the possibility of interference from third parties. As long ago as the closing years of the last century, David Johnson and David Post (1996) predicted that national regulators would have little success in controlling extraterritorial online activities, even though those activities have a local impact. While national regulators are not entirely powerless, the development of the Internet has dramatically changed the regulatory environment, creating new vulnerabilities to cybercrime and cyberthreats as well as new online suppliers, and community cultures. For local regulators, the question is how they can control access to drugs, or alcohol, or gambling, or direct-to-consumer genetic testing services, when Internet pharmacies, or online drink suppliers or casinos, or the like, all of which are hosted on servers that are located beyond the national borders, direct their goods and services at local regulatees.

Reacting to these challenges to Law 2.0 regulatory effectiveness, the thought occurs that social control might be more effective if new technologies were to be utilised as regulatory instruments. In particular, if human regulators were to be taken out of the equation, this might prevent corruption and capture, and if human regulatees had no practical option (because of technical measures) other than 'compliance,' then this might eliminate regulatee resistance. With these disruptive thoughts, we are on the cusp of Law 3.0.

Chapter 6

Law 2.0 and the 'crazy wall'

These days, no self-respecting TV crime drama is complete without its own 'crazy wall' (or 'evidence board'). Using a corkboard, a whiteboard, or a more sophisticated variation of the wall, the detectives who make up the investigatory team pin up photos of crime scenes, suspects, weapons, artefacts, and so on, scribble their questions and thoughts, and draw a bewildering set of lines that will help them establish the relevant connections before, in a lightbulb moment, figuring out who committed the offence.

In the same way, regulators addressing a new technology in a Law 2.0 frame might also start by using a crazy wall to focus their thoughts. As we have seen in the previous chapter, the challenge of getting the regulatory environment right is a multidimensional one; a Law 2.0 conversation can range across questions of legitimacy, effectiveness, and connection, and it is not just a matter of getting the rules right – the institutional apparatus that stands behind the rules must also be fit for purpose.

For regulators, it is the emergence of a new technology rather than the commission of a crime that starts the process. This is what prompts the scribbling. Suppose, for example, the emergent technology is augmented reality. Regulators might already have thought about the governance of virtual reality, but with augmented reality the anticipated benefits and risks might not be quite the same (Katell, Dechesne, Koops, and Meessen, 2019). With augmented reality, there might be beneficial applications in, say, the military and in policing, as well as in health and medicine. In those cases, regulators need to be confident that the further research and development of the technology is not impaired by limitations in the IP regime or by excessive *ex ante* safety checks or onerous *ex post* liability. On the other hand, the risks associated with augmented reality need to be assessed and managed.

When the regulators turn to risk assessment, they will have a list of, so to speak, the usual suspects. We do not want technologies that will kill or injure us or that will be harmful to the environment; we need to be careful about technologies that impinge on proprietary interests or that challenge our reasonable expectations of privacy; and at a more abstract level, we need to be sure that the applications of a particular technology are compatible with respect for human rights and human dignity. On the face of it, augmented reality might not seem to be particularly threatening or to present novel risks, but the experience with Google Glass as with Pokémania suggests otherwise. While the former represents a worrying extension of the data-collecting and behaviour-modifying power of the 'surveillance capitalists' (Zuboff, 2019: 155–157), the latter encourages game players to treat public spaces as gaming zones, raising issues about which kinds of activity may be pursued in particular spaces, about 'who gets to determine this, and [about] how technology may change this' (Katell, Dechesne, Koops, and Meessen at 289). Regulators need to heed the advice against legislating in haste only to repent at their leisure.

Or, again, suppose that it is 'deepfake' technology and its misleading images that is troubling regulators. Is there anything to write up on the benefit side of the crazy wall (say, in education or health) or are deepfakes simply an opportunity for making mischief, for portraying celebrities in an unflattering light, and for muddying the political waters by creating confusion about who actually said what? Arguably, though, the risks presented by deepfakes are systemic. If the root concern with augmented reality is that it presents agents with too much information or that it creates an informational asymmetry between agents (A knows all that there is to know about B, but B is not equally informed about A), the fundamental worry with deepfakes is that they compromise the informational ecosystem itself. As Al Gore put it in *The Assault on Reason* (2017), with the proliferation of 'fake news' – and, similarly, we can assume, with ubiquitous 'deepfakes' – the challenge is to restore 'a healthy information ecosystem that invites and supports the . . . essential processes of self-government in the age of the Internet so that [communities] can start making good decisions again' (294).

Having worked out what kind of regulatory scheme ideally needs to be put in place, regulators will turn once again to the wall as they consider how best to achieve their purposes. To what extent are hard-wired laws required? To what extent might self-regulation be more effective? Should there be a dedicated agency to monitor and police the use of augmented reality? There are many possible combinations of rules, many different institutional arrangements, and from the options presented on the wall, regulators must make their choice of approach.

Finally, even if a crime drama is usually over once the offender has been detected and convicted, the regulatory drama can continue. The use of augmented reality or deepfakes might evolve in ways that were not anticipated, the benefits and risks might not turn out in the ways that were expected, and (as we put it in the previous chapter) the regulation is no longer properly connected or sustainable. Once again, regulators will have to start their crazy wall to tackle the problem.

Chapter 7

Law 2.0 disrupted

Technology as a solution

In September 2009, a group of armed thieves, using a helicopter, carried out a spectacular raid on a cash depot in Stockholm (the 'Västberga heist,' as it became known). The owners of the depot offered a large reward for information leading to the arrest and conviction of the offenders or to the recovery of the stolen money. Just over a year later, a number of men were convicted of serious offences associated with the robbery, but the cash – millions of dollars of it – was not recovered. Although the Västberga story does not suggest that the criminal justice system in Sweden was ineffective or in a state of crisis, the reaction to the heist has been dramatic. In Sweden, the amount of cash in circulation has tumbled, and by 2014 only about 20% of retail payments were in cash. To be sure, this was not all down to the heist; Sweden's move towards a cashless economy has been aided and abetted by public concern about other high-profile robberies as well as by the commercial opportunities flowing from the development of new digital payment technologies (Heller, 2016).

For our purposes, what the aftermath of the Västberga heist illustrates is that a technological development (in cashless payment mechanisms) might be a solution to a particular regulatory problem (how to secure large amounts of cash). In other words, one of the disruptive effects of technological developments is to offer regulators options other than the use of rules. Such is the nature of the disruption to Law 2.0 that encourages the emergence of Law 3.0.

To be clear, the focus of the second disruption is not on the deficient content of prevailing legal rules, or on gaps, but on the availability of new technological instruments that can be applied for regulatory purposes. The response to such disruption is not that some rule changes or new rules are required but that the use of rules is not necessarily the most effective way of achieving the desired regulatory objective. Already, this presupposes a disruption to traditional patterns of legal thinking (a disruption to Law 1.0 and an evolution to Law 2.0) – that is to say, it presupposes a regulatory-instrumentalist and purposive mindset – and a willingness to think about turning to architecture, design, coding, AI, and the like as regulatory tools.

Arguably, we can find such a willingness as soon as people fit locks on their doors. However, as we have said, the disruption of Law 2.0 is predicated on both a variety and a sophistication of technological instruments – available to be employed not only for target-hardening purposes (like locks on doors) but also to exclude or disable potential wrongdoers as well as to immunise potential victims (for example, by automating processes) – that are strikingly different to the position in both pre-industrial and early industrial societies. If, in Law 2.0, advocates of 'smart regulation' (Gunningham and Grabosky, 1998) argue for the use of a complementary range of normative instruments, not relying on just one rule, in Law 3.0, smart regulation involves a dual and sustained focus on both an intelligent use of normative measures as well as appropriate design of places, products, and processes.

The take-up of technological tools can be charted on a spectrum running from soft to hard. At the soft end of the spectrum, the technologies are employed in support of the legal rules. For example, the use of surveillance technologies (such as CCTV) and/or identification technologies (such as DNA profiling, automatic number plate recognition of vehicles, or facial recognition) signals that rule-breaking is more likely to be detected; other things being equal, compliance with the rules is assisted and encouraged, but the strategy is still rule-based and the practical option of non-compliance remains. By contrast, at the hard end of the spectrum, the focus and the ambition are different. Here, measures of 'technological management' focus on limiting the practical (not the paper) options of regulatees, and whereas legal rules back their prescriptions with *ex post* penal, compensatory, or restorative measures, the focus of technological management is entirely *ex ante*, aiming to anticipate and prevent wrongdoing rather than punish or compensate after the event. As Lee Bygrave (2017) has put it in the context of the design of information systems and the protection of both intellectual property rights and privacy, the assumption is that, by embedding norms in the architecture, there is 'the promise of a significantly increased *ex ante* application of the norms and a corresponding reduction in relying on their application *ex post facto*' (at 755).

Taking up this last point, digital rights management is a clear example of the way in which manufacturers of digital products can code desired restrictions into the product itself. For instance, if a manufacturer of a DVD does not wish that DVD to be played anywhere in the world, but only in a certain region, the appropriate geo-restriction (or geo-blocking) can be designed into the product. Absent such a technical solution, the manufacturer might rely on restrictive terms and conditions in the contract or licence for use of the DVD but one would expect the *ex ante* technical solution to be more effective than *ex post* attempts to detect and enforce against breaches of the terms and conditions. With the technological disruption of Law 2.0, the turn to technical solutions by private parties is likely to provoke a similar mentality in public regulators, and vice versa.

This evolution in regulatory thinking is not surprising. Having recognised the limited fitness of traditional legal rules, and having taken a more regulatory approach, the next step surely is to think not just in terms of risk assessment and risk management but also to be mindful of the technological instruments that increasingly become available for use by regulators. In this way, the regulatory mindset is focused not only on the risks to be managed but also on how best to manage those risks (including making use of technological tools).

For example, with the development of computers and then the Internet and World Wide Web supporting a myriad of applications, it is clear that when individuals operate in online environments they are at risk in relation to both their 'privacy' and the fair processing of their personal data. Initially, regulators assumed that 'transactionalism' would suffice to protect individuals: in other words, it was assumed that, unless the relevant individuals agreed to, or consented to, the processing of their details, it would not be lawful. However, once it was evident that consumers in online environments routinely signalled their agreement or consent in a mechanical way, without doing so on a free and informed basis, a more robust risk-management approach invited consideration. Such an approach might still be rule-based, but the management might also be technological. In other words, once we are thinking about the protection of the autonomy of Internet-users or about the protection of their privacy, why not also consider the use of technological instruments in service of the regulatory objectives? As Woodrow Hartzog has argued, '[T]he design of popular technologies is critical to privacy, and the law should take it more seriously' (Hartzog, 2018: 7).

Indeed, in Europe, this is just what we find in the General Data Protection Regulation (GDPR) (EU Regulation 2016/679) and, similarly, in the new Copyright Directive (2019/790). While talk of 'privacy enhancing technologies' and 'privacy by design' has been around for some time, in the GDPR we see that this is more than talk. It is not just that the regulatory discourse is more technocratic; there are signs that this particular form of disruption is beginning to impact on regulatory practice. So in the GDPR, data controllers are required to take 'appropriate technical and organisational measures' to ensure that the requirements of the Regulation are met, and in Article 17 of the Directive, there is an implicit push towards the development of content recognition technologies that will assist cooperative arrangements between copyright holders and online service providers.

With initiatives of this kind, we can see that the second disruption is beginning to impact on regulatory practice. Granted, just how far this particular impact will penetrate remains to be seen. Nevertheless, the seeds of Law 3.0 have been sown. What we now have is the disruptive thought that technology might be part of the solution to our regulatory problems.

Chapter 8

Law 3.0

Coherentist, regulatory-instrumentalist, and technocratic conversations

Once upon a time, as railway trains packed with commuters pulled into Waterloo station, the doors were flung open, passengers hung out ready to jump off the train, and then duly disembarked from the moving train and hit the platform on the run. If this were happening in 2020, there would undoubtedly be a debate about the regulation of this dangerous practice. In 2020, health and safety standards would not tolerate such behaviour. If there were not already a rule prohibiting such conduct, there surely would be pressure to adopt one as a matter of urgency.

But, of course, in 2020, the trains that arrive at Waterloo station are of a quite different design to the steam trains and carriages of yesteryear. Today, not only are there far fewer doors on trains, all doors are centrally locked, and their unlocking is controlled by the train manager. It simply is not possible to dismount the train on the fly because the doors will not be unlocked until the train is at a standstill. In 2020, the design of the train takes care of health and safety; the design of the train takes the regulatory strain. In 2020, when commuters leave the train in the orderly way that they do, it is more a matter of what they can and cannot do than of what they ought or ought not to do. In 2020, the trains arriving at Waterloo station reflect the influence of Law 3.0 and the mindset that goes with it.

As we have said, the impact of the first wave of technological disruption is to destabilise the traditional coherentist mindset of Law 1.0 – the challenge comes from a Law 2.0 mindset that thinks in a more purposive regulatory-instrumentalist way. This disruptive effect is compounded by the second wave of disruption when regulatory-instrumentalism is taken in the more technocratic direction of Law 3.0. With each mindset, there are different questions that are focal, different framings, and different conversations that ensue.

In this chapter, three thumbnail sketches of these mindsets associated with Law 1.0, Law 2.0, and Law 3.0, respectively, are presented.

Coherentism

For present purposes, we can treat coherentist thinking as being defined by the following five, quintessentially Law 1.0, characteristics.

First, for coherentists, what matters above all is the integrity and internal consistency of legal doctrine. This is viewed as desirable in and of itself. Second, coherentists are not concerned with the fitness of the law for its regulatory purpose. Third, coherentists approach new technologies by asking how they fit within existing legal categories (and then try hard to fit them in) – for example, developments in biotechnology, in information technology, and in cryptotechnology, have raised (coherentist) questions about the way in which fundamental concepts and distinctions in property law map onto a range of 'things,' such as cell lines, gametes, personal data, and cryptoassets. Fourth, coherentists believe that legal reasoning should be anchored to guiding general principles of law. In the case of questions about the enforcement or non-enforcement of transactions, the foundational principles are that parties should be free to make their own deals and that it is the fact that parties have freely agreed to a deal that justifies holding them to the bargain. Fifth, coherentists assume that the function of private law, together with its guiding principles, is largely concerned with *ex post* correction and compensation.

It is worth lingering over the coherentist tendency to ask not whether the prevailing (and disrupted) rules are fit for purpose but how new phenomena can be fitted into traditional classification schemes or how they comport with general principles of law. Such a conservative mindset, it will be appreciated, mirrors that of the owners of BookWorld (see Chapter 2) in their initial reluctance to overhaul their classification scheme.

For coherentists, the focus is on the recognised legal concepts, categories, and classifications, and this is accompanied by a certain reluctance to abandon these concepts, categories, and classifications with a view to contemplating a bespoke response. For example, rather than recognise new types of intellectual property, coherentists will prefer to tweak existing laws of patents and copyright. Similarly, in transactions, coherentists (guided by existing law) will want to classify e-mails as either instantaneous or non-instantaneous forms of communication (or transmission), they will want to apply the standard contracts template to online shopping sites, they will want to draw on traditional notions of agency in order to engage electronic agents and smart machines, and they will want to classify individual 'prosumers' and 'hobbyists' who buy and sell on new platforms (such as platforms that support trade in 3D printed goods) as either business sellers or consumers. As the infrastructure for transactions becomes ever more technological, the tension between this strand of common law coherentism and regulatory-instrumentalism becomes all the more apparent. In sum, coherentism presupposes a world of, at most, leisurely change. It belongs to the age of the horse, not to the age of the autonomous vehicle.

Coherentism is, thus, the natural language of litigators and judges, who seek to apply the law in a principled way. It is also the default mode of thinking for many lawyers who take it that being trained 'to think like a lawyer' is synonymous with being trained to apply general principles of law to situations and phenomena both familiar and novel.

Regulatory-instrumentalism

In contrast with coherentism, we can treat regulatory-instrumentalism as being defined by the following six features.

First, regulatory-instrumentalism is not concerned with the internal consistency of legal doctrine. When regulatory-instrumentalists raise questions about consistency, they are typically making sure that particular regulatory interventions will complement others in serving specified regulatory objectives. Second, it is entirely focused on whether the law is instrumentally effective in serving specified regulatory purposes. Regulatory-instrumentalists do not ask whether the law is 'coherent' – other than in the sense of asking whether a group of related interventions pushes in the required regulatory direction – but whether it works. Third, regulatory-instrumentalism has no reservation about enacting new bespoke laws if this is an effective and efficient response to a question raised by new technologies. Fourth, the anchoring points for regulatory-instrumentalists are not the general principles that are established in the jurisprudence but current policy purposes and objectives. Fifth, alongside its policy focus, regulatory-instrumentalism in relation to new technologies tends to be orientated towards striking an acceptable balance between benefits and risks. Sixth, the tilt of the risk-management mindset that goes with regulatory-instrumentalism is towards *ex ante* prevention rather than *ex post* correction.

Regulatory-instrumentalism is, thus, the (democratically) mandated language of legislators, policy-makers, and regulatory agencies who talk the talk of Law 2.0. Conversely, while judges might have some responsibility for applying the spirit of policy-driven legislation, it is precisely the setting of regulatory policy that we think falls beyond the mandate of unaccountable judges.

Law 2.0 is relentlessly instrumentally rational. The question is: what works, what will serve certain specified purposes? When a regulatory intervention does not work, it is not enough to restore the status quo; instead, further regulatory measures should be taken, learning from previous experience, with a view to realising the regulatory purposes more effectively. Hence, the purpose of the criminal law is not simply to respond to wrongdoing (as corrective justice demands) but to reduce crime by adopting whatever measures of deterrence promise to work. Similarly, in a safety-conscious community, the purpose of tort law is not simply to respond to wrongdoing but to deter practices and acts where agents could easily avoid creating risks of injury and damage. For regulatory-instrumentalists, the path of the law should be progressive: we should be getting better at regulating crime and improving levels of safety.

According to Edward Rubin (2017), in the modern administrative state, the 'standard for judging the value of law is not whether it is coherent but rather whether it is effective, that is, effective in establishing and implementing the policy goals of the modern state' (at 328). Certainly, one of the striking features of the European Union has been the Law 2.0-spirit that has informed the single market project and that continues now in the digital Europe project. As the Commission (2015: 7) puts it:

> The pace of commercial and technological change due to digitalisation is very fast, not only in the EU, but worldwide. The EU needs to act now to ensure that business standards and consumer rights will be set according to common EU rules respecting a high-level of consumer protection and providing for a modern business friendly environment. It is of utmost necessity to create the framework allowing the benefits of digitalisation to materialise, so that EU businesses can become more competitive and consumers can have trust in high-level EU consumer protection standards. By acting now, the EU will set the policy trend and the standards according to which this important part of digitalisation will happen.

Post-Brexit, regulators in the United Kingdom will pursue similar objectives in their own way but, in both the EU and the UK, we can expect the coherence of the law with traditional principles to be very much a secondary concern.

Last but not least, it is characteristic of Law 2.0 that the thinking becomes much more risk-focused. In the criminal law and in torts, the risks that need to be assessed and managed relate primarily to physical and psychological injury and to damage to property and reputation; in contract law, it is economic risks that are relevant. So, for example, we see in the development of product liability a scheme of acceptable risk management that responds to the circulation of products (such as cars or new drugs) that are beneficial but also potentially dangerous. However, this response is still in the form of a revised *rule* (it is not yet technocratic), and it is still in the nature of an *ex post* correction (it is not yet *ex ante* preventive). Nevertheless, it is only a short step from here to a greater investment in *ex ante* regulatory checks (for food and drugs, chemicals, and so on) and to the use of new technologies as preventive regulatory instruments. In other words, it is only a short step from risk-managing regulatory-instrumentalist thinking to a more technocratic mindset.

Technocratic

The third mindset, evolving from a regulatory-instrumentalist view, is one that is distinctively technocratic. In response to the demand that 'there needs to be a law against this,' the technocratic mindset, rather than drafting new rules, looks for technological solutions. Such a mindset is nicely captured by Joshua Fairfield (2014: 39) when, writing in the context of non-negotiable terms and conditions in online consumer contracts, he remarks that 'if courts [or, we might say, the rules of contract law] will not protect consumers, robots will.'

Of course, in communities that are still committed to traditional values or to regulating by rules, there is likely to be some resistance to a technocratic mentality and to technical solutions. For example, in the United States, a proposal to design vehicles so that cars were simply immobilised if seat belts were not worn was eventually rejected. Although the (US) Department of Transportation estimated that the so-called interlock system would save 7,000 lives per annum and prevent 340,000 injuries, there was a popular pushback against this technological fix. Taking stock of the legislative debates of the time, Jerry Mashaw and David Harfst (1990: 140) remark:

> Safety was important, but it did not always trump liberty. [In the safety lobby's appeal to vaccines and guards on machines] the freedom fighters saw precisely the dangerous, progressive logic of regulation that they abhorred. The private passenger car was not a disease or a workplace, nor was it a common carrier. For Congress in 1974, it was a private space.

Today, similar debates might be had about the persistent and dangerous use of mobile phones by motorists. As we noted in Chapter 1, if we are to be technocratic in our approach, perhaps we might seek a design solution that disables phones within cars, or while the user is driving. However, once automated vehicles relieve 'drivers' of their safety responsibilities, it seems that the problem will drop away – rules that penalise humans who use their mobile phones while driving will become redundant; humans will simply be transported in vehicles and the one-time problem of driving while phoning will no longer be an issue.

With rapid developments in AI, machine learning, and blockchain, a question that will become increasingly important is whether (and if so, the extent to which) a community sees itself as distinguished by its commitment to governance by rule rather than by technological management. In some smaller-scale communities or self-regulating groups, there might be resistance to a technocratic approach because compliance that is guaranteed by technological means compromises the context for trust – this might be the position, for example, in some business communities (where self-enforcing transactional technologies are rejected). Or, again, a community might prefer to stick with regulation by rules because rules (unlike technological measures) allow for some interpretive flexibility, or because it values public participation in setting standards and is worried that this might be more difficult if the debate were to become technocratic.

While the contrast between a technocratic approach and coherentism is sharp – the former not being concerned with doctrinal integrity and not being entirely focused on restoring the status quo prior to wrongdoing – the contrast with regulatory-instrumentalism is more subtle. For both regulatory-instrumentalists and technocrats, the law is to be viewed in a purposive and policy-orientated way, and indeed, as we have said, the technocratic approach can be regarded as a natural evolution from regulatory-instrumentalism. In both mindsets, it is a matter of selecting the tools that will best serve desired purposes and policies, and so long as technologies are being employed as tools that are designed to assist with a rule-based regulatory enterprise – as is the case with the examples of drones at Gatwick Airport and harmful online content that we mentioned earlier in the book – the technocratic approach might be viewed as merely an offshoot from the stem of regulatory-instrumentalism.

That said, once technocrats contemplate interventions at the hard end of the spectrum, their thinking departs from order based on rules to one based on technological management, from correcting and penalising wrongdoing to preventing and precluding wrongdoing, and from reliance on rules and standards to employing technological solutions. At this point, Law 3.0 might itself be under pressure – or, at any rate, it might be under pressure if technical solutions and technological management are becoming the default strategy and not just something to be considered alongside a rule-based approach.

Chapter 9

Tech test case I

Liability for robot supervisors

Contemplating the potential liability of robots, or of those who use or rely on robots, John Frank Weaver (2014: 89) poses the following hypothetical:

> [S]uppose the Aeon babysitting robot at Fukuoka Lucle mall in Japan is responsibly watching a child, but the child still manages to run out of the child-care area and trip an elderly woman. Should the parent[s] be liable for that kid's intentional tort?

On these particular facts, especially the fact that the robot is 'responsibly' watching a child, the parents seem like the obvious defendants. However, if there was any suggestion that the robot might have been supervising the children less than 'responsibly,' might there then be a question about the liability of the robot itself or of its owners or controllers?

Consider, for example, the case of *Carmarthenshire County Council v Lewis* (1955), where a nursery schoolteacher supervising a three-year-old child was distracted when a second child fell and was hurt. While the teacher was attending to the second child, the first child wandered off out of the school building and through an open gate onto an adjoining road. In order to avoid hitting the child, a lorry driver swerved, collided with a telegraph pole, and was killed. The claimant was the lorry driver's widow. In these circumstances, the House of Lords (disagreeing with the lower courts) found that the teacher was not personally at fault – she had acted responsibly. However, it was held that the Council had failed to explain how the child had been able to get out of the school, and without such an explanation, it was to be inferred that the County Council had failed to take reasonable precautions to prevent an unattended child leaving the school and causing an accident of this kind.

Now, imagine that instead of a human schoolteacher supervising the children we have a robot looking after the children. On these facts, what would we say about the liability of the robot or of the County Council who rely on the robot to look after the young children at the nursery school? In the light of the discussion in the previous chapters, what we will say will depend on whether we are viewing the situation through the lens of Law 1.0, Law 2.0, or Law 3.0.

If we view the situation through the lens of Law 1.0, we will address it in a coherentist way. We will ask whether robot supervisors are analogous to human supervisors; can we treat robots as being 'personally' responsible or 'at fault'? If not, we have a problem. If robots are not to be treated as humans, how are we to treat them? As for the possible liability of the County Council, that is less problematic. As with the parents in Weaver's hypothetical, guided by principles of corrective justice, we will ask whether it would be fair, just and reasonable to hold the Council (or the parents at the mall) liable to compensate the injured parties.

If we view the situation through the lens of Law 2.0, our approach will be regulatory-instrumentalist. Here, the thinking would be that before retailers, such as the shop at the mall, are to be licensed to introduce robot babysitters, and parents permitted to make use of robocarers, there needs to be a collectively agreed scheme of compensation should something 'go wrong.' On this view, the responsibilities and liabilities of the parents would be determined by the agreed terms of the risk-management package. Similarly, a regulatory-instrumentalist would want to assess the risks and benefits of relying on robot supervisors in publicly funded schools and then determine what would be an acceptable balance of interests (including what would be an acceptable compensatory scheme in the event of an accident).

If we view the situation through the lens of Law 3.0, our approach will also be regulatory-instrumentalist but now with a technocratic dimension. In Law 3.0 we are thinking not only about the fitness of the rules but also about possible technical solutions. At the school, the technical solution might have been simply to keep the gate locked (there was considerable discussion about this in the case). At the mall, quite what measures of technological management might be suggested is anyone's guess – perhaps an invisible 'fence' at the edge of the care zone so that children simply could not stray beyond the limits. However, thinking about the puzzle in this way, the question would be entirely about designing the machines and the space in a way that collisions between children and mall-goers, or between schoolchildren and passing vehicles, could not happen.

Before we leave the scenario in *Carmarthenshire v Lewis*, imagine a further twist. Imagine that the lorry was not driven by a human but was fully autonomous and that it had been programmed to swerve if there was a risk of colliding with a human. Now, let us suppose, the claim for compensation is by the widow of a passenger who was in, but not driving, the vehicle. What would we say about such a situation?

For many, the principal point of discussion would be whether the vehicle should have been programmed to save the life of the child at the possible expense of any passengers in the vehicle. In fact, one of the main questions generated by the development of autonomous vehicles has been the ethics of dilemmas of just this kind. Let us suppose though that the community has debated the ethics, agreed on a standard design, and that the vehicle in question has the standard coding for situations of this kind. What can we now say about the question of legal liability?

According to Jonathan Morgan (2017: 537):

> From the industry perspective, liability is arguably the innovative manufac-
> turer's greatest concern. The clashing interests raise in acute form the classic
> dilemma for tort and technology: how to reconcile reduction in the number of
> accidents (deterrence) and compensation of the injured with the encourage-
> ment of socially beneficial innovation? Not surprisingly there have been calls
> for stricter liability (to serve the former goals), and for immunities (to foster
> innovation). But in the absence of any radical legislative reform, the existing
> principles of tort will apply – if only faute de mieux.

This is surely right. As we have said, if the regulation of an emerging technology is presented in a legislative setting, a risk-management approach is likely to
prevail, with regulators trying to accommodate inter alia the interest in beneficial innovation together with the interest in risks being managed at an acceptable
level. It will be a Law 2.0 conversation. On the other hand, if the question arises
in a court setting, it is more likely that both litigants and judges will talk the coherentist talk of negligence and fault even though, as Morgan observes, the common
law technique of reasoning by analogy via existing categories, far from being
'common sense,' is 'obfuscatory' (539).

Picking up on these reflections, the first thing that we might say is that it is little
wonder that we find the law relating to autonomous vehicles puzzling when we
try, in a coherentist Law 1.0 way, to apply the principles for judging the negligence
of human drivers to questions of liability concerning vehicles in which there is no
human in control. Perhaps we are asking ourselves the wrong questions.

Second, if liability questions are taken up in the courts, judges (reasoning like
coherentists) will try to apply notions of a reasonable standard of care to responsibility for very complex technological failures. Just imagine, for example, the
difficulties faced by a judge trying to figure out whether, following the two tragic
crashes involving Boeing 737 max planes, there was any failure of care on the
part of the pilots (in their interaction with the technology) or in the design of
the unique MCAS automated flight control feature fitted to the planes. In some
cases, judges might or might not treat following the standard design as sufficient,
but where the technology is at the leading edge, there might be no 'standard' and
there would need to be an inquiry into whether the manufacturers were acting
'responsibly' – to say the least, not an easy question.

Third, if regulators in a legislative setting approach the question of liability
and compensation with a risk-management mindset, they will not need to chase
after questions of fault. Rather, the challenge will be to articulate the most acceptable (and financially workable) compensatory arrangements that accommodate
the interests in transport innovation and the safety of passengers and pedestrians.
For example, one proposal (advanced by Tracy Pearl, 2018) is that there should be
an autonomous vehicle crash victim compensation fund to be financed by a sales
tax on such vehicles. Of course, as with any such no-fault compensation scheme,

much of the devil is in the detail – for example, there are important questions to be settled about the level of compensation, whether the option of pursuing a tort claim remains open to victims, and what kind of injuries or losses are covered by the scheme.

Fourth, as Morgan says, the better way of determining the liability arrangements for autonomous vehicles is not by litigation but 'for regulators to make the relevant choices of public policy openly after suitable democratic discussion of which robotics applications to allow and which to stimulate, which applications to discourage and which to prohibit' (539).

Putting all this more generally, we can say that it is important that communities ask themselves the right questions, understanding, first, that coherentist framings are very different to regulatory-instrumentalist ones, and second, that the availability of technological solutions invites reflection (as is characteristic of Law 3.0), not only about adjustments that might be made to the liability rules but also about whether such rules are really necessary at all.

Chapter 10

Tech test case II

Smart shops, code law, and contract law

When we were discussing BookWorld (in Chapter 2), we mentioned the (also fictitious) Fully Automated Bookstore (FAB). At FAB, an array of technologies is embedded in the architecture of the store so that when customers enter, they are identified, the books that they select are itemised and displayed on the customer's basket, and at the point of exit the price of the books is debited to the customer's account. The process is seamless, the technology invisible, and performance is assured. What role or relevance, we might wonder, do the rules of the law of contract – rules about the formation of contracts, the terms and conditions of contracts, breaches of contract, and remedies – have to the transactions processed at FAB?

If customers who wish to buy books at FAB first have to enter into a master agreement, in which the terms and conditions (the rules) for the use of FAB are specified, then we do not need to look for a contract each time a customer buys a book at the store; instead, it is the master agreement that is the critical contract. In the event that there is a mistake or a problem arising from a purchase, the rights and remedies of the parties will be governed by the master contract. The law of contract still has a handle on FAB transactions but not in the way that we have hitherto applied it.

That said, if there is no such master agreement, or if FAB's terms and conditions or the automated bookstores' codes of practice do not adequately protect the interests of consumers, there might need to be a bespoke regulatory intervention, a legislative framework that sets out the standard rules for transactions of this kind. Although contract law is no longer the governing instrument, the code of law (albeit in a Law 2.0 regulatory form) still governs transactions at FAB. We might not have the rule of contract law, but the Rule of Law more generally still obtains.

With some technologies, however, we might anticipate that the code of law is challenged by the operation of a technology. For example, the code of intellectual property law might be challenged by the technologies employed for the purposes of digital rights management (see Chapter 7). Here, the coding of digital products might reflect a rebalancing of the interests of IP rights-holders with those of end users of the products, such rebalancing being in favour of the former and over and above what is treated as acceptable by the legal code. Similarly, the thought occurs that blockchain-based smart contracts might be coded to generate outcomes (or payments) that the law of contract (or any other part of the legal code) would *not* generate. In other words, the thought occurs that there might be two universes, two logics, two codes, running parallel and potentially coming into tension with one another. This is the possibility to be considered in this chapter.

An uneasy coexistence

In an insightful discussion of blockchain and its many applications, Karen Yeung (2019) highlights three ways in which the law community (and the Rule of Law) might relate to the blockchain community (and the rule of code). First, where the blockchain community challenges the authority of the law community, we can expect the relationship to be adversarial as the latter strives to assert its sovereignty; the pressure will be for technological code to yield to legal code. Second, where the technology seeks to complement or supplement the law, the relationship should be positive and supportive. Third, where the technology represents an alternative to the law (being, as it were, 'competitive' but not conflictual as such) the relationship is more complex and less predictable. In cases of this third kind, Yeung suggests that the relationship is likely to be one of 'uneasy coexistence,' characterised by an attitude of 'mutual suspicion' (210).

This prospect of an uneasy coexistence between the legal community and the blockchain community raises questions about the relationship between the law of contract and automated transactional technologies (including so-called smart contract applications that run on blockchain). On the face of it, where such transactional technologies are employed, the relationship with the law will be of the third kind; it will be one of uneasy coexistence. For these are not technologies that are designed directly to challenge the authority of the law or simply to complement existing laws but rather to give transacting parties an enforcement option that is an alternative to taking court action.

Nevertheless, to the extent that the legal community (guided by the law of contract) and the blockchain community (guided by smart contract coding) operate with their own logics, there is the possibility that there might be a lack of congruence between the legal and the technological view of the transaction. In the event of such a lack of congruence, the question is whether it would be regarded as problematic, if so why, and then how a court (viewing the problem in a Law 1.0 way) might respond.

A lack of congruence

Putting the matter somewhat schematically, we can say that from a Law 1.0 perspective, some examples of a lack of congruence might not be seen as problematic in themselves, some might be seen as arguably problematic, and some might be seen as clearly problematic. How a court then responds is another question.

For instance, where the law of contract has restrictive rules in relation to the enforcement of claims by third-party beneficiaries, a smart contract (or some other automated process) might achieve a transfer of value to a third party with which the law of contract (and the courts) would not assist. Or, again, a smart contract might be coded for particular remedial payments in ways that are different to the law of contract's default position on fair and appropriate compensatory awards of damages (see Chapter 3). In such cases, if a party were to invite a court to bring the technological effects into line with the principles of contract law (to restore congruence), a court might decline to do so either because it does not regard the deviation as problematic in itself or because it treats the parties as having elected to subject their dealings to a different (but permissible) technological code.

On the other hand, some cases of a lack of congruence would almost certainly be seen as problematic, particularly where the technological effect is seen as contrary to public policy. For example, it has long been a principle of the common law of contract that provisions that are 'penal' in nature should not be enforced. Following the Supreme Court's decision in *Cavendish Square Holding* (2015), the critical question is whether the party relying on the allegedly penal clause has a legitimate interest in doing so and whether the provision is proportionate relative to that interest. Applying this test to provisions in standard fiat contracts is not entirely straightforward, and where smart contract provisions are tied to a cryptocurrency which (as has been the case with Bitcoin) is subject to extraordinary fluctuations, the application of the legal test might be even less straightforward. Nevertheless, it would be surprising if a court declined to adjudicate where a smart contract effect was challenged as being 'penal'; and if the court took the view that the effect was indeed penal, then it would be equally surprising if it did not make an appropriate order disallowing the payment.

Public policy also drives the regulatory thinking that underlies the modern law of consumer contracts. In this context, consider how a court might respond to a scenario in which the technology disables the use of goods or services where a consumer is in breach of some restriction in the contract (e.g., relating to the purposes for which the goods or services might be used or where or when they might be used). In such circumstances, how might a court respond to a challenge (brought by an individual consumer or by a representative body) where the complaint rested on the unfair terms provisions in the Consumer Rights Act 2015? While a coherentist-minded court might be tempted to dismiss the case on the ground that the transaction does not qualify as a 'contract' within the meaning of the CRA, it is more likely that a court would see this as a problematic lack of congruence. This leads to the question of whether it is public policy to protect consumers who participate in all permitted transactions with traders regardless of whether the transaction is offline, online, smart, or otherwise. If so, the protections in the CRA should be applied. Moreover, even if there is no example of an unfair term in the statutory Schedule that squarely fits the case, the asymmetric nature of the technological effect, coupled with its potential unfairness and its lack of proportionality, will probably persuade a court that it needs to take protective action (compare Durovic and Janssen, 2019).

Perhaps, the clearest example of a problematic lack of congruence is where the technology facilitates illegal transactions. Famously, in *Holman v Johnson* (1775), Lord Mansfield – writing long before transactional technologies had presented questions of non-congruence – remarked that 'No court will lend its aid to a man who founds his cause of action upon an immoral or an illegal act' (343). Taking this principle at face value, we would expect a court, at the very least, to decline to assist a litigant who is party to a transaction that is a means to an immoral or illegal end, and we would expect this to be the case whether the transaction is a fiat contract or a smart contract.

For example, suppose that A employs B, a contract killer, to kill C for an agreed fee. The contract is committed to a technology that not only hides the identities of A and B but also automatically transfers the agreed fee to B once C has been killed. I take it that such an outcome would not be sanctioned by the law of contract and that the courts should (at minimum) do nothing to encourage such an illegal activity. However, what if the technology ensures that B is paid before B kills C but then B fails to kill C? Should a court assist A to recover the payment made to B? As a matter of general principle, coherentists might think that the courts should do nothing to assist either A or B. However, in the recent case law – notably in *Patel v Mirza* (2016) and in *DC Merwestone* (2016) – we find the Supreme Court taking a somewhat nuanced approach. According to the Court, there are seriously (and intrinsically) illegal acts/transactions and there are merely illegal acts/transactions, and there are some lies that are more material than others. Instead of a general rule, we have a shift to a case-by-case examination which is sensitive to principles of fairness and proportionality – but an approach, nevertheless, that is still essentially coherentist in nature.

Arguably, though, reliance on coherentist principles is not the right way to tackle the questions to which illegal contract cases can give rise. Rather, our approach should be more regulatory, more Law 2.0. For example, imagine that a legislative provision that is designed to protect tenants against rapacious landlords makes it a criminal offence for a landlord to collect 'ancillary' charge payments from tenants. The legislation, however, is silent on the question of whether agreements between landlords and tenants concerning the now prohibited charges are legally enforceable. Clearly, the statutory silence notwithstanding, no court would contemplate ordering a tenant to make a prohibited payment to the landlord, and it would be equally out of line with the regulatory policy if a court declined to assist a tenant to recover a prohibited payment to a landlord for the reason that this would be to aid a party to an illegal contract.

In *Patel v Mirza*, Lord Neuberger – whose judgement we can treat as being 'representative' of the sweep of judicial views in this case – hints at a more regulatory approach by noting on several occasions that the relevant law is 'based on policy' (para 161). It is not entirely clear, however, whose policy it is – judicial or legislative – or which policy it is that is being relied on. In particular, it seems to be assumed that, in general, it is good policy to return the parties to an illegal transaction to the position that they were in *ex ante*. In favour of such a policy, it can be said that there is no reason why one party to such a transaction should gain at the expense of the other. Yet, if these parties are complicit in a prohibited transaction, why should the law take an interest in correcting any inter-party unfairness? Regardless of whether smart contracts are involved, why not simply treat the parties as acting entirely at their own risk if they engage in such dealings?

What should we conclude about non-congruence? First, where the law of contract is itself in a state of flux, it might not be clear whether the technological effect is out of line with (non-congruent with) the code of law. Second, where the technological effect clearly is non-congruent, it might not be viewed as problematic, and the courts might decline to assist a party who invites them to restore congruence. Third, even where the technological effect is clearly non-congruent and problematic, it remains to be seen how far coherentist-minded courts will accept a role in enforcing public policy. As for the uneasy coexistence itself, we should remind ourselves that whatever uneasy coexistence there might be between legal effects (as in Law 1.0) and technological effects, these are more directly resolved as a matter of policy in Law 2.0, and in Law 3.0 they are brought together and 'internalised' by a technocratic mindset.

Chapter 11

Easterbrook and the
Law of the Horse

Nearly 25 years ago, a group of 'cyberlaw' enthusiasts met at a conference in Chicago. This was one of the first conferences of its kind. The conference is recalled, however, not for the innovative contributions of these cyberlawyers but for an unexpectedly conservative intervention made by Judge Frank Easterbrook (1996).

Arguing that 'the best way to learn the law applicable to specialized endeavors is to study general rules' (207), Easterbrook claimed that a course on the 'Law of Cyberspace' would be as misconceived and unilluminating as a course on 'The Law of the Horse.' It would be 'shallow' and it would 'miss unifying principles' (207). Rather, the better approach, Easterbrook contended, was

> to take courses in property, torts, commercial transactions, and the like. . . . [For only] by putting the law of the horse in the context of broader rules about commercial endeavors could one really understand the *law* about horses.
>
> (208)

Nevertheless, the law of cyberspace was a horse that was destined to bolt. Easterbrook's doubts notwithstanding, courses and texts on 'cyberlaw,' or 'Internet law,' or 'the law of e-commerce,' or the like, abound, and few would deny that they have intellectual integrity and make pedagogic sense. Similarly, research centres that are dedicated to the study of cyberlaw (or law and technology more generally) have mushroomed and are seen as being in the vanguard of legal scholarship.

That said, was Easterbrook wrong? And, if so, why exactly was he wrong? In the light of what has been said about the evolution of Law 1.0, through Law 2.0, and to Law 3.0, it becomes clear that Easterbrook's argument is predicated on Law 1.0 being the operative paradigm. However, it was not the operative paradigm in 1996 and it certainly is not the operative paradigm today.

In 1996, cyberlaw was still in its formative stages and there were a number of questions of law that were being tested out in the courts – for example, about the liability of parties for defamatory online content and about conflicts of laws matters concerning the applicable law and the jurisdiction of the courts. Indeed, one of the early cases on notice of terms in relation to 'shrink wrap' licences and software contracts, *ProCD v Zeidenberg* (1996), was heard by Easterbrook himself. To this extent, the conversation was, in line with Easterbrook's position, about the application of the existing general principles of the relevant areas of law. A discrete 'Law of Cyberspace' was yet to emerge. It was not too long, though, before bespoke laws for e-commerce, for Internet content (and the liability of Internet service providers or other intermediaries), for cybercrime, for online privacy, and so on were being put in place (conspicuously so in Europe). Such laws were quite distinct from the general principles of law that Easterbrook had in mind, and they were shaped by regulatory-instrumentalist thinking. In short, Easterbrook's remarks seem to assume that the laws governing cyberspace would be Law 1.0 when it rapidly became clear that the relevant laws were largely Law 2.0.

Twenty-five years on from Easterbrook's intervention, it seems that he failed to grasp the extent of the technological disruption of the law. Easterbrook not only failed to anticipate the regulatory thinking of much of the incoming law of cyberspace but he failed to anticipate that the law might engage with cybertechnologies as both regulatory targets (technologies to be regulated) and regulatory tools (technologies to be used by regulators). In other words, Easterbrook did not foresee Law 2.0 on the near horizon and he did not foresee the emergence of Law 3.0 on the mid horizon.

With the benefit of hindsight, it is easy to be wise. Nevertheless, Easterbrook was wrong, and in retrospect his mistake was not so much to default to a coherentist mindset but to underestimate the disruptive effects of technology on the law and at the same time to over-estimate the flexibility of the general principles (here we might also question the assessment of the Jurisdiction Taskforce to which we referred in Chapter 1). If we are to 'really understand the *law*,' it is essential to step outside a Law 1.0 mindset. Only then is it possible to recognise the extent of the disruption wrought by new technologies and, concomitantly, the significance of legal order.

In sum, the problem with Easterbrook's approach is that it is a denial of (or in denial about) disruption. While this might be appropriate in the age of the horse, it is not at all appropriate in an age of disruptive cybertechnologies. In such an age, we need to reimagine the field of legal interest.

Part two

Law reimagined

Chapter 12

Law as one element in the regulatory environment

If Law 3.0 is our regulatory future, then we need to press the reset button. We need to reimagine law. As a first step, we should broaden the field for juristic inquiry by operating with a notion of the regulatory environment that accommodates both normative rule-based and non-normative technocratic approaches. Critically, we must correct a fixation with Law 1.0 by reimagining law within a regulatory environment that is no longer limited to guidance given by rules or norms. In other words, we have to make space for the technocratic dimension of Law 3.0.

What would such a regulatory environment look like? Famously, Clifford Shearing and Phillip Stenning (1985) highlighted the way in which, at Disney World, the vehicles that carry visitors between locations act as barriers (restricting access). However, theme parks are no longer a special case. We find similar regulatory environments in many everyday settings, where along with familiar laws, rules, and regulations, there are the signs of technological management – for example, we find mixed environments of this kind in homes and offices where air-conditioning and lighting operate automatically, in hotels where the accommodation levels can only be reached by using an elevator (and where the elevators cannot be used and the rooms cannot be accessed without the use of security key cards), and perhaps *par excellence* in what geographers refer to as the 'code/ space' that we find at airports.

Entering a modern terminal building, while there are many airport rules to be observed – for example, regulations concerning parking vehicles, smoking in the building, or leaving bags unattended, and so on – there is also a distinctive architecture that creates a physical track leading from arrival and check-in to departures and boarding. Along this track, there is nowadays an 'immigration and security zone,' dense with identifying and surveillance technologies, through which passengers have little choice other than to pass. Moreover, if we ever have the misfortune to reach the departure lounge but then find that there is no plane to board, we will soon realise that there is no simple track that will allow us to retrace our steps back to the arrivals area and exit the building: the pathway at the airport is designed to be one-way only, taking passengers from arrivals to departure, not the other way round. In this conjunction of architecture and surveillance and identifying technologies, we have the non-normative dimensions of the airport's regulatory environment – the fact of the matter is that, if we wish to board our plane, we have no practical option other than to follow the technologically managed track.

Similarly, if we want to shop at an Amazon Go store, we have no choice other than to subject ourselves to the technologically managed environment of such stores, and of course, if we visit Amazon or any other platform online, we will probably do so subject to both the specified terms and conditions for access and whatever technological features are embedded in the site. Needless to say, this is a far cry from the Law 1.0 questions about contractual offer and acceptance that continue to puzzle us in relation to the layout of those early self-service stores that departed from the traditional across-the-counter model.

If we treat the regulatory environment as essentially a signalling and steering environment, then each such environment operates with a distinctive set of regulatory signals that are designed to channel the conduct of regulatees within, so to speak, a regulated sphere of possibility. Of course, one of the benefits of technologies is that they can expand our possibilities; without aircraft, we could not fly. Characteristically, though, the kind of technological management that we are contemplating is one that restricts or reduces existing human possibilities (albeit, in some cases, by way of a trade-off for new possibilities). In other words, while normative regulation is directed at actions that are possible – and that remain possible – technological management engages with spheres of possibility but in ways that restructure those regulatory spaces and redefine what is and is not possible.

It is important to understand that the technological measures contemplated in Law 3.0 are varied. They include the design of products (such as the golf carts that we will meet in Chapter 14, or computer hardware and software, or digital payments in place of cash) and processes (such as the automated production and driving of vehicles, or the provision of consumer goods and services), the design of places (such as the Metro, or theme parks and airports) and spaces (particularly online spaces), and (in the future) the design of people. Typically, such measures are employed with a view to managing certain kinds of risks by excluding (a) the possibility of certain actions which, in the absence of this strategy, might be subject only to rule regulation, or (b) human agents who otherwise might be implicated (whether as rule-breakers or as the innocent victims of rule-breaking) in the regulated activities. Moreover, technological management might be employed by both public regulators and by private self-regulating agents (such as corporations protecting their IP rights or supermarkets protecting their merchandise and their trolleys).

Schematically, where the use of technological management is available as a regulatory option, the process can be presented in the following terms:

- Let us suppose that a regulator, R, has a view about whether regulatees should be required to do x, permitted to do x, or prohibited from doing x (the underlying normative view).

- R's view could be expressed in the form of a rule that requires, permits, or prohibits the doing of x (the underlying rule).
- But R uses (or directs others to use) technological management rather than a rule.
- And R's intention in doing so is to translate the underlying normative view into a practical design that ensures that regulatees do or do not do x (according to the underlying rule).
- The ensuing outcome is that regulatees find themselves in environments where the immediate signals relate to what can and cannot be done, to possibilities and impossibilities, rather than to the underlying normative pattern of what ought or ought not to be done.

This description, however, is highly schematic, and what such a process actually amounts to in practice – in particular, how transparent the process is, how much debate there is about the underlying normative view and then about the use of technological measures – will vary from one context to another, from public to private regulators, between one public regulator and another, and between one private regulator and another.

It also should be emphasised that the ambition of hard technical measures is to replace the rules by controlling the practical options that are open to regulatees. In other words, technological management goes beyond technological assistance in support of the rules. Of course, regulators might first turn to technological instruments that operate in support of the rules. For example, in an attempt to discourage shoplifting, regulators might require or encourage retailers to install surveillance and identification technologies or technologies that sound an alarm should a person carry goods that have not been paid for through the exit gates. However, this is not yet full-scale technological management. Once technological management is in operation, shoppers will find that it is simply not possible to take away goods without having paid for them.

Finally, to avoid any misunderstanding, it should be said that the rule-based (normative) dimension of the regulatory environment should be broad and inclusive. The normative context for law goes far beyond the formal rules of law (see Chapter 25), this being reflected in a large literature on the coexistence of, and interaction between, formal legal rules and informal norms and codes (so to speak, the 'living law'). We certainly do not want to subtract any of this. Rather, what Law 3.0 reflects is the use of technological tools for regulatory purposes and the coexistence of rule and non-rule instruments not only in the regulatory toolbox but also in the everyday experience of regulatees. Accordingly, we need to conceive of the regulatory environment as comprising both formal and informal normative codes as well as non-normative technological tools and codes. Law 3.0, is more than a particular technocratic mode of reasoning, it is also a state of coexistent codes and conversations.

Chapter 13

Mapping the regulatory environment

Recalling our fictitious bookshop, BookWorld, the pressure to reimagine the business format is largely economic: unless BookWorld makes some changes, it will no longer be competitive. Why, though, should jurists reimagine law? If their interests are purely doctrinal, if their mindset is purely Law 1.0, jurists can continue to engage with their traditional puzzles and lines of inquiry. However, to the extent that technological management displaces rules as the regulatory instrument of choice, traditional legal scholarship loses its relevance. Like those who are experts in a language that is no longer spoken, coherentist lawyers (following Easterbrook) will be experts in a form of social ordering or disputing that is no longer practised. Moreover, if jurists hope to be able to contribute to debates about the legitimacy of particular forms of social ordering or particular exercises of power, they need to think beyond the coherentism of Law 1.0, and they need to reimagine law as one element in a larger configuration of power.

A general map

We can concede that jurists will have different cognitive interests and priorities. Nevertheless, we can propose two sets of features that would give shape to a very general map of the reimagined field. First, the map should indicate which type of measures or instruments are being used, and second, it should indicate whether the source of the measure or instrument is public (and typically top-down) or private (and often bottom-up).

Employing the first indicator, the map should tell us whether a particular regulatory environment, or a particular regulatory space, is constituted by rules or by non-rule technologies (or, indeed, by some combination of rules and non-rule technologies). Where we are in zones that are regulated by rules, we are in familiar territory; we have centuries of jurisprudential reflection to help us. However, where non-rule technologies are in play, it is a very different story. As Sheila Jasanoff (2016: 9–10) has remarked, even though

> technological systems rival legal constitutions in their power to order and
> govern society . . . there is no systematic body of thought, comparable to

centuries of legal and political theory, to articulate the principles by which technologies are empowered to rule us.

Accordingly, once we have our most general map in place, we can begin work on a map that will aid our reimagination of law specifically where non-rule technologies are in play.

Our general map should also tell us whether the source of the measure is public (and typically top-down) or private (and often bottom-up) – in other words, whether the regulator is public or private. In much traditional legal scholarship, the focus is on rules that have been promulgated by public law-making bodies. As critics of this approach have objected, this focus neglects the rule-making activities of private bodies. However, even with an expanded focus, we are still presupposing that we are operating in rule-governed zones. Once we move into regulatory spaces where non-rule technologies apply, then we are in largely uncharted territory. Even so, it would be surprising if we did not think it important to know whether these technologies have been initiated and are being controlled by public or by private regulators.

That said, it must be admitted that the distinction between public and private is notoriously contestable and that the distinction between top-down and bottom-up regulation is both crude and far from exhaustive. For example, top-down government regulators might enlist the aid of non-governmental intermediaries (such as Internet service providers or platform providers) or they might adopt a co-regulatory approach setting general targets or objectives for regulatees but leaving them to determine how best to comply. With new technologies occupying and disrupting regulatory spaces, regulators need to reimagine how best to regulate. As Albert Lin (2018: 965) says in his analysis of new distributed innovative technologies (such as DIYbio, 3D printing, and the platforms of the share economy), these new forms of dynamic activity 'confound conventional regulation.' In response, Lin argues, it turns out that '[g]overnance of distributed innovation . . . must be both distributed and innovative' (1011). There is no one-size-fits-all, and the regulatory environment that is most acceptable and effective is likely to have elements of both top-down and bottom-up approaches together with elements that fit neither of these types.

Nevertheless, as a first cut at reimagining regulatory spaces, we can work along two axes. On one axis it is the balance between reliance on rules and reliance on technologies that is indicated, and on the other axis it is the extent to which regulatory interventions are public and/or top-down or private and/or bottom-up that is indicated.

A specific mapping of technological measures

Once we are in areas that are regulated by non-rule technological measures, how should we get our bearings? I suggest that our map should indicate, first, what the nature of the particular measure is (specifically where it lies on a spectrum between soft and hard intervention) and second, the locus of the intervention (specifically where it lies on a spectrum between external [to human agents] and internal [to human agents]).

With regard to the first indicator, we can differentiate between, on the one hand, those technological measures that are merely supportive of existing rules or assistive or advisory in relation to decision-making, and, on the other, measures of technological management proper that aim to eliminate or redefine some part of an agent's practical options. For example, the use of surveillance and identification technologies in the criminal justice system may simply support the rules of the criminal law; and the use of use of AI in police practice and in criminal justice decision-making may be simply assistive and advisory (compare Chapter 21). By contrast, if vehicles cannot be driven unless seat belts are engaged, we have full-scale technological management.

Some years ago, Mireille Hildebrandt (2008) drew a distinction between 'regulative' and 'constitutive' technological features. Whereas the former are in the nature of assistive or advisory technological applications, the latter represent full-scale technological management. By way of an illustrative example, Hildebrandt invites readers to imagine a home that is enabled with a smart energy meter:

> One could imagine a smart home that automatically reduces the consumption of energy after a certain threshold has been reached, switching off lights in empty rooms and/or blocking the use of the washing machine for the rest of the day. This intervention [which is a case of a 'constitutive' technological intervention] may have been designed by the national or municipal legislator or by government agencies involved in environmental protection and implemented by the company that supplies the electricity. Alternatively [this being a case of a 'regulative' technological intervention], the user may be empowered to program her smart house in such a way. Another possibility [again, a case of a 'regulative' technological intervention] would be to have a smart home that is infested with real-time displays that inform the occupants about the amount of energy they are consuming while cooking, reading, having a shower, heating the house, keeping the fridge in function or mowing the lawn. This will allow the inhabitants to become aware of their energy consumption in a very practical way, giving them a chance to change their habits while having real-time access to the increasing eco-efficiency of their behaviour.

(174)

Similarly, Pat O'Malley (2013: 280) charts the different degrees of technological control on a spectrum running from 'soft' to 'hard' by reference to the regulation of the speed of motor vehicles:

> In the 'soft' versions of such technologies, a warning device advises drivers they are exceeding the speed limit or are approaching changed traffic

regulatory conditions, but there are progressively more aggressive versions. If the driver ignores warnings, data – which include calculations of the excess speed at any moment, and the distance over which such speeding occurred (which may be considered an additional risk factor and *thus* an aggravation of the offence) – can be transmitted directly to a central registry. Finally, in a move that makes the leap from perfect detection to perfect prevention, the vehicle can be disabled or speed limits can be imposed by remote modulation of the braking system or accelerator.

Accordingly, whether we are considering smart cars, smart homes, or smart regulatory styles, we need to be sensitive to the way in which the regulatory environment engages with regulatees, whether it directs signals at regulatees enjoining them to act in particular ways, or whether the technology of regulation simply imposes a pattern of conduct upon regulatees irrespective of whether they would otherwise choose to act in the way that the technology now dictates.

At all points on this spectrum, whether the technological instrument is simply advisory and assistive, or becomes a 'nudge' (again running from soft to hard), or becomes a full-blown measure of technological management, we need to be sensitised to the significance of the particular nature of the technological measure.

This takes us to the second specific indicator, the locus of the intervention. For the most part, our assumption is that technological instruments are being embedded in places and spaces in which human agents find themselves or with which they interact. Hence, we can talk about technologically managed zones or zones that are rule-governed. However, the proliferation of smart portable or wearable devices, together with many other smart products (such as autonomous vehicles) suggests that the relevant regulatory technological features are not so much zones into which human agents enter but extensions of the human agent. Nevertheless, we might persist with the idea that such technological instruments are still external to the agent. However, with the development of various kinds of augmented reality and implants, the line between external and internal locations becomes more difficult to maintain. As Franklin Foer (2017: 2) has suggested, the development of wearables such as 'Google Glass and the Apple Watch [might] prefigure the day when these companies implant their artificial intelligence within our bodies.' In due course, if, in addition to coded spaces and coded products, we have coded human agents (analogous to coded robots), the line between external and internal signalling would have been crossed.

Taking stock, our general map will enable us to identify the type of regulatory measure (rule or non-rule technological) employed together with the source of that measure (public or private), and where the measure is non-rule technological, our specific map will enable us to identify whether it is a soft or hard intervention and whether the locus is external or internal. Even if we are not quite sure how to respond to a particular measure, this initial mapping at least helps us to reconstruct our sense of the landscape of Law 3.0 and to indicate where we stand in any particular case.

Chapter 14

The complexion of the regulatory environment

Imagine a fictitious golf club, 'Westways.' The story at Westways begins when some of the older members propose that a couple of carts should be acquired for use by members who otherwise have problems in getting from tee to green. There are sufficient funds to make the purchase but the greenkeeper expresses a concern that the carts might cause damage to Westways' carefully manicured greens. The proposers share the greenkeeper's concerns and everyone is anxious to avoid causing such damage. Happily, this is easily solved. The proposers, who include most of the potential users of the carts, act in a way that is respectful of the interests of all club members; they try to do the right thing, and this includes using the carts in a responsible fashion, keeping them well clear of the greens.

For a time, the carts are used without any problem. However, as the membership of Westways changes – particularly as the older members leave – there are some incidents of irresponsible cart use. The greenkeeper of the day suggests that the club needs to take a firmer stance. In due course, the club adopts a rule that prohibits taking carts onto the greens and that penalises members who break the rule. Unfortunately, this intervention does not help; indeed, if anything, the new rule aggravates the situation. While the rule is not intended to license irresponsible use of the carts (on payment of a fine), this is how some members perceive it, and the effect is to weaken the original 'moral' pressure to respect the interests of fellow members of the club. Moreover, members know that, in some of the more remote parts of the course, there is little chance of rule-breakers being detected.

Taking a further step to discourage breaches of the rule, it is decided to install a few CCTV cameras around the course at Westways. However, not only is the coverage patchy (so that it is still relatively easy to break the rule without being seen in some parts of the course) but old Joe, who is employed to watch the monitors at the surveillance control centre, is easily distracted. Members soon learn that he can be persuaded to turn a blind eye in return for the price of a couple of beers. Once again, the club fails to find an effective way of channelling the conduct of members so that the carts are used in a responsible fashion.

It is at this juncture that the club turns to a technological fix. The carts are modified so that, if a member tries to take the cart too close to one of the greens (or to take the cart off the course) they are warned and, if the warnings are ignored, the cart is immobilised. At last, thanks to technological management, the club succeeds in realising the benefits of the carts while also protecting its greens.

As we trace the particular history at our fictitious club, Westways, we see that the story starts with an informal 'moral' understanding. In effect, just as in the early days of eBay, regulation rests on the so-called Golden Rule: that is to say, the rule is that members should use the carts (or the auction site) in the way that they would wish others to use them. It then tries to reinforce the moral signal with a rule (akin to a law) that sends a prudential signal (namely, that it is in the interests of members to comply with the rule lest they incur the penalty). However, the combination of a prudential signal with a moral signal is not altogether a happy one because the former interferes with the latter. When CCTV cameras are installed, the prudential signals are amplified to the point that they are probably the dominant (but still not fully effective) signals. Finally, with technological management, the signals change into a completely different mode: once the carts are redesigned, it is no longer for members to decide on either moral or prudential grounds to use the carts responsibly. At the end of the story, the carts cannot be driven onto the greens, and the signals are entirely to do with what is possible and impossible.

What the story at Westways illustrates is the significant changes that take place in the 'complexion' of the regulatory environment; with each regulatory initiative, the 'signalling register' changes from moral, to prudential, and then to what is possible. With each move, the moral register is pushed further into the background. Similarly, while the signals associated with Law 1.0 and Law 2.0 are both moral and prudential, the technological measures of Law 3.0 signal what is possible (or not possible).

In this chapter, we can begin to equip ourselves to think more clearly about the changing complexion of the regulatory environment – and particularly the changes introduced by technological management. Starting with the idea of the 'regulatory registers' (that is, the ways in which regulators seek to engage the practical reason of their regulatees) and their relevance, we find that technological management relies on neither of the traditional normative registers (moral and prudential). With technological management, there is a dramatic change in the complexion of the regulatory environment, and the question is whether the use of such a regulatory strategy can be justified. For example, if technical measures are more effective than rules in preventing drones from endangering the safety of aircraft, does it matter so much that the use of these measures changes the complexion of the regulatory environment?

The regulatory registers

By a 'regulatory register,' I mean the kind of signal that regulators employ in communicating with regulatees. There are three such registers, each of which represents a particular way in which regulators attempt to engage the practical reason (in the broad and inclusive sense of an agent's reasons for action) of regulatees. Thus:

1   in the moral register, the coding signals that some act, x, categorically ought or ought not to be done relative to standards of right action – regulators thus signal to regulatees that x is, or is not, the right thing to do;
2   in the prudential register, the coding signals that some act, x, ought or ought not to be done relative to the prudential interests of regulatees – regulators thus signal to regulatees that x is, or is not, in their (regulatees') self-interest; and
3   in the register of practicability or possibility, the environment is designed in such a way that it is not reasonably practicable (or even possible) to do some act, x – in which case, regulatees reason, not that x ought not to be done but that x cannot be done.

In an exclusively moral environment, the primary normative signal (in the sense of the reason for the norm) is always moral, but the secondary signal, depending upon the nature of the sanction, might be more prudential. In traditional criminal law environments, the signals are more complex. While the primary normative signal to regulatees can be either moral (the particular act should not be done because this would be immoral) or paternalistically prudential (the act should not be done because it is contrary to the interests of the regulatee), the secondary signal represented by the deterrent threat of punishment is prudential. As the regulatory environment relies more on technological instruments, the strength and significance of the moral signal fades; here, the signals to regulatees tend to accentuate that the doing of a particular act is contrary to the interests of regulatees, or that it is not reasonably practicable, or even that it is simply not possible.

The relevance of the registers

How does a framing of this kind assist our inquiries? Crucially, if we share the aspiration for regulatory interventions that are both effective and legitimate, then the framing of our inquiries by reference to the regulatory registers draws our attention to any technology-induced drift from the first to the second register (where moral reasons give way to prudence) and then from the second to the third register (where the signal is no longer normative). Sophisticated technologies of control surely will appeal to future classes of regulators in both the public and the private sectors, but communities with moral aspirations need to stay alert to the corrosion of the regulatory conditions that give meaning to their way of life. The key point is that in moral communities the aspiration is to do the right thing, respecting persons is doing the right thing, and it matters to moral communities that such respect is shown freely and for the right reason – not because we know that we are being observed or because we have no practical option other than the one that the regulators have left for us. Quite simply, even if the technology channels regulatees towards right action, the technologically secured pattern of right action is not at all the same as freely opting to do the right thing. One agent might be protected from the potentially harmful acts of others, but moral virtue, as Ian Kerr (2010) protests, cannot be automated.

While surveillance technologies, such as the use of CCTV at Westways, leave it to regulatees to decide how to act, technologies can be used in ways that harden the regulatory environment, by designing out or disallowing the choice that regulatees previously had – precisely in the way that the problem is eventually solved at the golf club. As we have said, such hard technologies speak only to what can and cannot be done, not to what ought or ought not to be done; these are non-normative regulatory interventions.

It might well be that our concerns about technological management change from one context to another. It might be that moral concerns about the use of technological management are at their most acute where the target is *intentional* wrongdoing or *deliberately* harmful acts – even though, of course, some might argue that this is exactly where a community most urgently needs to adopt (rather than to eschew) technological management. That said, it surely cannot be right to condemn all applications of technological management as illegitimate. For example, should we object to modern transport systems on the ground that they incorporate safety features that are intended to design-out the possibility of human error or carelessness (as well as intentionally malign acts)? Or should we object to the proposal that we might turn to the use of regulating technologies to replace a failed normative strategy for securing the safety of patients who are taking medicines or being treated in hospitals?

Recognising the significance of a shift from rules to technical measures, a moral community will be greatly concerned about technological management that prevents regulatees from doing the right thing (rather than forcing them to do it) or that forces them to do the wrong thing. Scaling this up, is there a risk that technological management might compromise the possibility of engaging in responsible moral citizenship?

Recalling the case of Rosa Parks, a black woman who refused to move from the 'white-only' section of a bus, Evgeny Morozov (2013) points out that this important act of civil disobedience was possible only because

> the bus and the sociotechnological system in which it operated were terribly inefficient. The bus driver asked Parks to move only because he couldn't anticipate how many people would need to be seated in the white-only section at the front; as the bus got full, the driver had to adjust the sections in real time, and Parks happened to be sitting in an area that suddenly became 'white-only.'
>
> (204)

However, if the bus and the bus stops had been technologically enabled, this situation simply would not have arisen – Parks would either have been denied entry to the bus or she would have been sitting in the allocated section for black people. Morozov continues:

> Will this new transportation system be convenient? Sure. Will it give us Rosa Parks? Probably not, because she would never have gotten to the front of the bus to begin with. The odds are that a perfectly efficient seat-distribution system – abetted by ubiquitous technology, sensors, and facial recognition – would have robbed us of one of the proudest moments in American history. Laws that are enforced by appealing to our moral or prudential registers leave just enough space for friction; friction breeds tension, tension creates conflict, and conflict produces change. In contrast, when laws are enforced through the technological register, there's little space for friction and tension – and quite likely for change.
>
> (205)

In short, technological management disrupts the assumption made by liberal legal theorists who count on acts of direct civil disobedience being available as an expression of responsible moral citizenship.

Suppose that legislation is introduced that specifically authorises or mandates the use of a suite of smart technologies on and around buses in order to maintain a system of racial segregation on public transport. Those who believe that the legislative policy is immoral might have opportunities to protest before the leg-islation is enacted; they might be able, post-legislation, to demonstrate at sites where the technology is being installed; they might be able to engage in direct acts of civil disobedience by interfering with the technology; and they might have opportunities for indirect acts of civil disobedience (breaking some other law in order to protest about the policy of racial segregation on public transport). Regulators might then respond in various ways – for example, by creating new criminal offences that are targeted at those who try to design round technologi-cal management. Putting this more generally, technological management might not altogether eliminate the possibility of principled moral protest. The particular technology might not always be counter-technology proof and there might remain opportunities for civil disobedients to express their opposition to the background regulatory purposes indirectly by breaking anti-circumvention laws, or by engag-ing in strategies of 'data obfuscation,' or by initiating well-publicised 'hacks,' or 'denial-of-service' attacks or their analogues.

Nevertheless, if the general effect of technological management is to squeeze the opportunities for traditional direct acts of civil disobedience, ways need to be found to compensate for any resulting diminution in responsible moral citizen-ship. By the time that technological management is in place, for many this will be too late; for most citizens, symbolic and evocative expressions of conscientious objection and non-compliance will no longer be an option. This suggests that the compensating adjustment needs to be *ex ante*: that is to say, it suggests that respon-sible moral citizens need to be able to air their objections before technological management has been authorised for a particular purpose; and, what is more, the opportunity needs to be there to challenge both an immoral regulatory purpose and the use of (morality-corroding) technological management.

Taking stock, whether we are talking about individual moral lives or moral community, some precaution is in order. If we knew just how much space a moral community needs to safeguard against the automation of virtue, we might be able to draw some regulatory red lines. However, without knowing these things, a precautionary approach (according and protecting a generous margin of operational space for moral reflection, for moral reason, and for moral objection) looks prudent. Without knowing these things, the cumulative effect of adopting technological management – at any rate, in relation to intentional wrongdoing – needs to be a standing item on the regulatory agenda.

Chapter 15

Law 3.0 and liberty

The pianos at St Pancras

In the impressive concourse at the renovated St Pancras railway station, there are a couple of upright pianos. There seem to be no restrictions on who can play or what can be played, and most days a diverse range of piano music can be heard being played by a motley set of players. In what is already a bustling place of many sounds, the pianists add to the metropolitan mix.

Suppose, though, that a significant number of people who travel through St Pancras would prefer the volume to be turned down somewhat. They complain that the pianists add to the sounds of the station but not in a good way. Responding to such complaints and guided by a Law 3.0 approach, the station managers would consider, first, whether some rules might be introduced to regulate the playing of the pianos (for example, restricting the times when the pianos may be played) and, second, whether some non-rule regulatory measures might be taken (for example, removing one or more of the pianos).

Now, the point of this short chapter is not to rehearse what has already been said but to put the spotlight on the two dimensions of a regulated space (such as the concourse at St Pancras railway station) and the different foci of, respectively, rules and technological measures. As a species, humans have certain capabilities which, given an appropriate space or place, they can exercise. For example, in the station concourse, they can walk and talk, and if a piano is free, they can sit and play some music. So far, so unregulated. However, if these activities are to be regulated by rules, the restrictions imposed by the rules overlay the particular space (or, the particular sphere of possibility). The focus for the rules is on what is regulated, not on the capabilities of humans and not on what *can* be done in the regulated space. By contrast, where technological measures are employed, the focus is on resculpting human capabilities or the features of the regulatory space (as by removing the pianos) precisely in order to regulate what can be done.

This difference in focus is one of the key things that sets Law 3.0 apart from Law 1.0 and Law 2.0, both of which are focused on applying or making changes to the regulatory code rather than making changes to the regulatory space, or the sphere of possibility, itself. Moreover, this difference is extremely important for our appreciation of how Law 3.0's technological measures can impact on our individual liberty (Brownsword, 2017).

Mainstream thinking about liberty values a rule framework that gives us options – for example, a rule that makes it entirely optional whether one plays one of the pianos at St Pancras station. Given this rule, those who opt to play do no wrong to anyone, and those who opt not to play equally do no wrong. According to the rules, playing is optional; there is a liberty to play or not to play. However, this view of liberty is somewhat limited because it does not speak to whether the option is actually available in practice. According to the rule, playing a piano at St Pancras station is optional whether or not there are pianos actually standing in the concourse. Taking a limited view, there would be neither more nor less liberty to play a piano at the station irrespective of whether there were pianos available to be played and irrespective of whether pianos were being installed or being removed.

While, in a Law 1.0 or a Law 2.0 conversation, questions about liberty focus on the relevant rules and whether or not conduct is optional, in a Law 3.0 conversation, questions about liberty need also to focus on whether regulators are restricting the practical options that we have. Indeed, the more that regulators rely on technological measures, the more that it will be the impact on our practical liberty that needs to be monitored. In rule-governed situations, it will be important to ask whether the rules treat some conduct as optional. In situations where technological measures are employed, however, the important question will be whether that same conduct is something that in practice is actually at our option – in other words, is this something that we actually can choose to do or not to do?

When the men in overalls remove the pianos from the concourse at St Pancras railway station, we know that the world has changed. While whatever the rules provide about piano playing at the station might not have changed, playing a piano there is no longer a practical option. To this extent, our liberty, particularly the liberty of prospective piano players, has been diminished. With Law 3.0, the risk is not so much that our paper liberties might be diminished by visible and well-advertised changes to the rules but that our practical liberties might be much less openly and transparently reduced by the overnight removal of the pianos. This is not to say that the expansion or contraction of our paper (rule-based) liberties is no longer relevant. It is to say instead that, with the coming of Law 3.0, we should be alert to monitor and debate the impact on our practical liberty of the increasingly technological mediation of our transactions and interactions coupled with the use of technological management for regulatory purposes.

In sum, in Law 3.0, if we value our liberty, we should pay more attention to what we can and cannot do and be somewhat less concerned about what the rules say we ought and ought not do.

Chapter 16

Law 3.0

The thin end of the wedge, and the thick end

There is a view that in the not too distant future the functions of law will be discharged by smart technologies, that smart machines will serve as legal functionaries. For example, it has been claimed that recent developments in artificial intelligence (AI) and machine learning (ML) foreshadow a transition to a 'legal singularity' – a singularity where there will only rarely be disputes about the legal significance of agreed facts (Alarie, 2016).

The context for Alarie's discussion is one in which we want to know how a particular legal distinction (such as that between 'an employee' and 'an independent contractor') will be applied to a given set of facts. While there are many reported decisions concerning the application of the distinction, each decision hinges on the particular facts, as a result of which it is not easy to be confident about the application of the distinction to a new set of facts. The idea in Alarie is that an AI tool could be trained using the reported decisions to predict the outcome of new cases and in due course even to decide those cases. How should we respond?

Perhaps we should simply 'be philosophical.' If AI and ML are the thin end of a technological wedge that will transform the practice of law, then so be it. As Anthony Quinton famously remarked – at a time when it was proposed that the rules at New College, Oxford, should be relaxed to permit undergraduates to sleep with women undisturbed at weekends – if we are faced with the thin end of a wedge then 'better [that] than the other.'

Or perhaps we should be sceptical: AI and ML might be the thin end of a wedge, but at least in relation to common law adjudication, the wedge is unlikely to get much thicker. In this context, as Christopher Markou and Simon Deakin (2019) have argued, the practical utility of AI tools might be limited by their backward-looking nature as well as their unwelcome lock-in effects. Moreover, as others have cautioned, we should not underrate the essentially human and social elements in adjudication (see, e.g., Crootof, 2019).

A further response might be to push back and resist. Strikingly, in France, section 33 of the Justice Reform Act 2019 provides that '[t]he identity data of magistrates and members of the judiciary cannot be reused with the purpose or effect of evaluating, analysing, comparing or predicting their actual or alleged professional practices.' Even though those who commit the new offence can face a custodial

sentence of up to five years, the extent to which the mere redaction of the identifying data will impede the use of AI and ML to predict the outcome of cases is
unclear. Nevertheless, section 33 represents a significant expression of resistance.

Given the bigger picture perspective of Law 3.0, a rather different response
might be made. This is that the focus on using AI tools to give guidance on the
application of legal rules or principles together with the fear that smart machines
might replace litigation lawyers and judges betrays a fixation with Law 1.0. Scep-
ticism about the idea that smart machines might undertake the coherentist reasoning that is characteristic of Law 1.0 betrays not just a human-centric view but a
Law 1.0 mindset. Breaking free from the grip of Law 1.0, we will see that the
discourse of Law 3.0 is already engaging with the prospect of smart machines
undertaking regulatory functions, channelling and rechannelling human conduct.
If we think of technology as a wedge under the idea of law as the enterprise of
subjecting human conduct to the governance of rules, then it is already happening.
Moreover, to put the point provocatively, the technological wedges into the channelling function of law are going in thick end first.

As for the idea of a legal singularity, we should say that fully automated adjudication, taking the place of a Law 1.0 conversation is not enough. Before we
have legal singularity, AI and other technologies have to take over the kind of
channelling function that is central to Law 2.0, and of course the interest that we
find distinctively in Law 3.0, in building both enabling and disabling features into
regulatory spaces, is the beginning of such a process. In other words, although
the adoption of AI in dispute-avoidance and dispute-settlement is a thin end of a
technological wedge, it is the early signs of technological management in Law 3.0
that is more significant because the thick end of the wedge is precisely that the
regulatory task of channelling and rechannelling conduct is carried out by the
technologies rather than by the rules. If and when human conduct is subjected to
the governance of smart machines rather than rules, we have singularity.

At all events, if we are to respond to the wedge represented by technological
measures, and (at the thickest end) by technological management, then this will
take a radical rebooting of our legal thinking, starting with our understanding of
regulatory responsibilities, and then reshaping the Rule of Law and our conception of coherence in the law. It is to these urgently needed adjustments that we turn
in the next part of the book.

Part three

Living with Law 3.0

Chapter 17

The benchmarks of legitimacy

The range of regulatory responsibilities

If the direction of travel (from Law 1.0 to Law 3.0) is towards a more regulatory approach, with the focus being on what works and how best to serve specified policies, there is a risk that instrumental considerations come to dominate. As Robert Merton put it so eloquently in his Foreword to Jacques Ellul's *The Technological Society* (1964: vi), we need to beware civilisations and technocrats that are 'committed to the quest for continually improved means to carelessly examined ends.'

While in Law 1.0 it might be right to say that the general principles are either unexamined or that they are simply assumed to be legitimate, no such exemption from scrutiny, no such assumption, is made in Law 2.0 or Law 3.0. Generally speaking, the legitimacy of the regulators' purposes or policies is a matter for active debate – or, at any rate, this is so where we are dealing with public regulators. In other words, even if Merton's warning might be appropriate in relation to the activities of some private regulators, it is not yet an urgent concern in relation to public regulators. Nevertheless, there are two striking problems in relation to our public discourse concerning regulatory responsibilities. The first is that we assume that whatever particular principles or purposes we take to be guiding, they are in the final analysis reasonably and rationally contestable; the second is that we engage in all manner of balancing exercises (between rights, interests, public policy, and so on) without any clear sense of there being a hierarchy that guides deciding between conflicting considerations. In short, we lack foundations, and we lack hierarchy. At least, at our fictitious bookshop BookWorld, the owners knew that it would be unthinkable to close the brick-and-mortar store and move the business online. Accordingly, a priority for living with Law 3.0 is to restore some order to our understanding of regulatory responsibilities.

In that spirit, I suggest that we frame our thinking by articulating three tiers of regulatory responsibility, the first tier being foundational, and the responsibilities being ranked in three tiers of importance. At the first and most important tier, regulators have a 'stewardship' responsibility for maintaining the preconditions for human social existence, for any kind of human social community. I will call these conditions 'the commons.' At the second tier, regulators have a responsibility to respect the fundamental values of a particular human social community, that is to say, the values that give that community its particular identity. At the third tier, regulators have a responsibility to seek out an acceptable balance of legitimate interests. The responsibilities at the first tier are cosmopolitan and non-negotiable (the red lines here are hard); the responsibilities at the second and third tiers are contingent, depending on the fundamental values and the interests recognised in each particular community. Any conflicts between these responsibilities are to be resolved by reference to the tiers of importance: responsibilities in a higher tier always outrank those in a lower tier.

Given that the regulatory instruments available in Law 3.0 promise to be more effective than ever, we need to be clear about which regulatory purposes and positions are legitimate. It is the scheme of regulatory responsibilities that provides the benchmarks for legitimacy. In what follows, I will speak briefly to each of the three tiers of that scheme.

The regulatory responsibility for the commons

It is an article of faith in the medical profession that doctors should, first, do no harm (to their patients). For regulators, the equivalent injunction should be, first, to ensure that no harm is done to the generic conditions that underpin the lives and prospects of their regulatees.

This injunction rests on a simple but fundamental idea. This is that it is characteristic of human agents that, as *humans*, they have certain biological needs (the need for a range of life-supporting conditions) and that, as *agents*, they have the capacity to pursue various projects and plans whether as individuals, in partnerships, in groups, or in whole communities. Sometimes, the various projects and plans that they pursue will be harmonious, but often human agents will find themselves in conflict or competition with one another as their preferences, projects, and plans clash. However, before we get to particular projects or plans, before we get to conflict or competition, there needs to be a context in which the exercise of agency is possible. This context is not one that privileges a particular articulation of agency; it is prior to, and entirely neutral between, the particular plans and projects that agents individually favour. The conditions that make up this context are generic to agency itself. In other words, there is a deep and fundamental critical infrastructure, a commons, for any community of agents. It follows that any agent, reflecting on the antecedent and essential nature of the commons, must regard the critical infrastructural conditions as special. From any practical viewpoint, prudential or moral, that of regulator or regulatee, the protection of the commons must be the highest priority. Indeed, this is so obvious that we really should not need striking schoolchildren and Swedish teenagers to remind us of such self-evident truths.

Accordingly, we expect regulators to be mindful that we, as humans, have certain biological needs and that first, there should be no encouragement for technologies that are dangerous in that they compromise the conditions for our very existence. Second, given that we have a (self-interested) sense of which technological developments we would regard as beneficial, we will press regulators to support and prioritise such developments – and, conversely, to reject developments that we judge to be contrary to our self-interest. Third, even where proposed technological developments are neither dangerous nor lacking utility, some will argue that they should be prohibited (or at least not encouraged) because their development would be immoral.

If we build on this analysis, we will argue that the paramount responsibility for regulators (whether they otherwise think like coherentists, regulatory-instrumentalists, or technocrats) is to protect and preserve:

- the essential conditions for human existence (given human biological needs);
- the generic conditions for human agency and self-development; and
- the essential conditions for the development and practice of moral agency.

These, it bears repeating, are imperatives for regulators in all regulatory spaces, whether international or national, public or private. Of course, determining the nature of these conditions will not be a mechanical process, and I do not assume that it will be without its points of controversy. Nevertheless, let me give an indication of how I would understand the distinctive contribution of each segment of the commons.

In the first instance, regulators should take steps to protect and preserve the natural ecosystem for human life. At minimum, this entails that the physical well-being of humans must be secured; humans need oxygen, they need food and water, they need shelter, they need protection against contagious diseases, if they are sick they need whatever medical treatment is available, and they need to be protected against assaults by other humans or non-human beings. It follows that the intentional violation of such conditions is a crime against, not just the individual humans who are directly affected, but humanity itself.

Second, the conditions for meaningful self-development and agency need to be constructed: there needs to be a sufficient sense of self and of self-esteem, as well as sufficient trust and confidence in one's fellow agents, together with sufficient predictability to plan, so as to operate in a way that is interactive and purposeful rather than merely defensive. Let me suggest that the distinctive capacities of prospective agents include being able to form a sense of what is in one's own *self*-interest; to choose one's own ends, goals, purposes, and so on ('to do one's own thing'); and to form a sense of one's own identity ('to be one's own person'). With a supportive set of generic conditions, human life becomes an opportunity for agents to be who they want to be, to have the projects that they want to have, to form the relationships that they want, to pursue the interests that they choose to have, and so on. In the twenty-first century, no other view of human potential and aspiration is plausible; in the twenty-first century, it is axiomatic that humans are prospective agents and that agents need to be free.

The gist of these agency conditions is nicely expressed in a paper from the Royal Society and British Academy (2016: 5) where, in a discussion of data governance and privacy, we read:

Future concerns will likely relate to the freedom and capacity to create conditions in which we can flourish as individuals; governance will determine the social, political, legal and moral infrastructure that gives each person a sphere of protection through which they can explore who they are, with whom they want to relate and how they want to understand themselves, free from intrusion or limitation of choice.

In this light, we can readily appreciate that – unlike, say, Margaret Atwood's post-apocalyptic dystopia *Oryx and Crake* – what is dystopian about George Orwell's *1984* and Aldous Huxley's *Brave New World* is not that human *existence* is compromised but that human *agency* is compromised. In Frank Pasquale's (2015: 52) words, we can appreciate, too, that today's dataveillance practices, as much as *1984*'s surveillance, 'may be doing less to deter destructive acts than [slowly to narrow] the range of tolerable thought and behaviour.'

Third, the commons must secure the conditions for an aspirant moral community, whether the particular community is guided by teleological or deontological standards, by rights or by duties, by communitarian or liberal or libertarian values, by virtue ethics, and so on. The generic context for moral community is impartial between competing moral visions, values, and ideals, but it must be conducive to 'moral' development and 'moral' agency in a formal sense. So, for example, in her discussion of techno-moral virtues, (sous)surveillance, and moral nudges, Shannon Vallor (2016: 203) is rightly concerned that any employment of digital technologies to foster prosocial behaviour should respect the importance of conduct remaining 'our *own conscious activity and achievement* rather than passive, unthinking submission.' Echoing our remarks in Chapter 14, Vallor argues that, unless we can explain *why* we act in good ways, why the ways we act *are* good, or *what* the good life for a human being or community might be, we cannot treat a person as a moral being. Quite simply, there is a risk that moral agency is compromised by technologies that do too much regulatory work.

Agents who reason impartially will understand that each human agent is a stakeholder in the commons where this represents the essential conditions for human existence together with the generic conditions of both self-regarding and other-regarding agency, and it will be understood that these conditions must, therefore, be respected. While respect for the commons' conditions is binding on all human agents, it should be emphasised that these conditions do not rule out the possibility of prudential or moral pluralism. Rather, the commons represents the preconditions for both individual self-development and community debate, giving each agent the opportunity to develop his or her own view of what is prudent as well as what should be morally prohibited, permitted, or required. However, the articulation and contestation of both individual and collective perspectives (like all other human social acts, activities, and practices) are predicated on the existence of the commons.

The regulatory responsibility to respect the community's fundamental values

Beyond the fundamental stewardship responsibilities, regulators are also responsible for ensuring that the fundamental values of their particular community are respected. Just as each individual human agent has the capacity to develop their own distinctive identity, the same is true if we scale this up to communities of human agents. There are common needs and interests but also distinctive identities.

From the middle of the twentieth century, many nation states have expressed their fundamental (constitutional) values in terms of respect for human rights and human dignity (Brownsword, 2014). These values clearly intersect with the commons' conditions, and there is much to debate about the nature of this relationship and the extent of any overlap – for example, if we understand the root idea of human dignity in terms of humans having the capacity freely to do the right thing for the right reason, then human dignity reaches directly to the commons' conditions for moral agency. However, those nation states that articulate their particular identities by the way in which they interpret their commitment to respect for human dignity are far from homogeneous. Whereas in some communities the emphasis of human dignity is on individual empowerment and autonomy, in others it is on constraints relating to the sanctity, non-commercialisation, non-commodification, and non-instrumentalisation of human life. These differences in emphasis mean that communities articulate in very different ways on a range of beginning of life and end of life questions as well as questions of human enhancement and so on.

As we noted in Chapter 8, one question to be addressed is whether, and if so how far, a community sees itself as distinguished by its commitment to regulation by rule. In some smaller-scale communities or self-regulating groups, there might be resistance to a technocratic approach. However, where a community is happy to rely on technological features rather than rules, it then has to decide whether it is also happy for humans to be out of the loop. Where the technologies involve AI, the 'computer loop' might be the only loop that there is. As Shawn Bayern and his co-authors (2017: 156) note, this raises an urgent question, namely: '[D]o we need to define essential tasks of the state that must be fulfilled by human beings under all circumstances?' Furthermore, once a community is asking itself such questions, it will need to clarify its understanding of the relationship between humans and robots – in particular, whether it treats robots as having moral status, or legal personality, and the like.

In Europe, the latter question is still under relatively early discussion with a number of views being expressed. However, in relation to the former question, Article 22 of the GDPR stakes out a default prohibition on solely automated decisions which have legal or other significant effects in relation to an individual, and it then provides for humans to be brought back into the loop where the default does not apply (see further Chapter 21).

It is, of course, essential that the fundamental values to which a particular community commits itself are consistent with (or cohere with) the commons' conditions, and if we are to talk about a new form of coherentism – as I will suggest we should in Chapter 22 – it should be focused in the first instance on ensuring that regulatory operations are so consistent.

The regulatory responsibility to seek an acceptable balance of interests

This takes us to the third tier of regulatory responsibility. As we have said, with the development of a regulatory-instrumentalist mindset, we find that much of traditional tort and contract law is overtaken by an approach that seeks to promote general policy objectives (such as supporting and encouraging beneficial innovation) while balancing this with countervailing interests. Given that different balances will appeal to different interest groups, finding an acceptable balance is a major challenge for regulators.

Today we have the perfect example of this challenge in the debate about the liability (both criminal and civil) of Internet intermediaries for the unlawful content that they carry or host. Should intermediaries be required to monitor content or simply act after the event by taking down offending content? In principle, we might argue that such intermediaries should be held strictly liable for any or some classes of illegal content, that they should be liable if they fail to take reasonable care, or that they should be immunised against liability even though the content is illegal. If we take a position at the strict liability end of the range, we might worry that the liability regime is too burdensome to intermediaries and that online services will not expand in the way that we hope, but if we take a position at the immunity end of the range, we might worry that this treats the Internet as an exception to the Rule of Law and is an open invitation for the illegal activities of copyright infringers, paedophiles, terrorists, and so on. In practice, most legal systems balance these interests by taking a position that confers an immunity, but only so long as the intermediaries do not have knowledge or notice of the illegal content. Predictably, now that the leading intermediaries are large US corporations with deep pockets, and not fledgling start-ups, many think that the time is ripe for the balance to be reviewed. However, finding a balance that is generally acceptable, in both principle and practice, is another matter.

While finding an acceptable balance might be messy and provisional, and while it might be the best that we can do at the time, we should not make the mistake of thinking that balancing is always the best that we can do. Where the stewardship of the commons or the values that define a particular community are at stake, there are clear priorities, and regulators have very different responsibilities.

Summing up, we might say that new technologies, whether as regulatory tools or as regulatory targets, should not be used unless they have a triple licence: a commons (tier one) licence, a community (tier two) licence, and a social (tier three) licence (see further Chapter 20).

Chapter 18

Uncertainty, precaution, stewardship

One of the characteristics of Law 3.0 is that it views technical measures and technology as a part of the regulatory solution. However, technology is also a problem and a challenge for regulators. In the early days of the development and application of a new technology, regulators will be uncertain about what risks it might present, and where a technology is widely adopted and perceived to be beneficial, it might then be too late for regulators to make effective interventions to mitigate risks. Recall, for example, the case of the Internet (arguably now running out of control and too much relied upon to be closed down) and the question of the responsibilities of online intermediaries with which we closed the previous chapter.

Emerging technologies give rise to many kinds of concern: one is that the application of a particular technology might present risks to human health and safety or to the environment (as is the case, for example, with much of the concern about both synthetic biology and nanotechnologies). Another concern is that the technology might be applied in ways that are harmful to fundamental values (as is the case with much human biotechnology and neurotechnologies as well as with information technologies where interests in privacy and confidentiality, and the like, are recurrent concerns). Those who harbour such concerns demand that regulators take protective action. However, in many cases, the context in which such demands are made is both deeply contested and clouded by uncertainty. For example, it might be uncertain which types of impact (whether an impact on human health, on the environment, on human rights, or human dignity, or whatever) a particular technology might have, or how likely it is that an impact will eventuate or whether, indeed, an impact (such as the destruction of human embryos) involves any kind of moral harm.

In such a context, how should regulators respond to calls for action? Quite reasonably, it might be suggested that regulators should strive to maintain a responsible and rational approach. As the Appellate Body at the WTO put it in the *Hormones* (1998: para 124) dispute, 'responsible, representative governments commonly act [and should act] from perspectives of prudence and precaution where risks of irreversible, e.g., life-terminating, damage to human health are concerned.' However, precautionary approaches are frequently accused of being irrational because they focus in a one-eyed way on the need to avoid a particular set of adverse consequences at the expense of considering the probability (or improbability) of the consequences actually eventuating (as well as ignoring the adverse consequences of making a precautionary intervention) (see, e.g., Sunstein, 2005). From a Law 3.0 perspective, informed by an appreciation of the full span of regulatory responsibilities, what should we make of this?

Precaution and scientific uncertainty

According to Principle 15 of the Rio Declaration of the United Nations Conference on Environment and Development (1992) (arguably, this being the foremost articulation of the Precautionary Principle, PP):

> In order to protect the environment, the precautionary approach shall be widely applied by States according to their capabilities. Where there are threats of serious or irreversible damage, lack of full scientific certainty shall not be used as a reason for postponing cost-effective measures to prevent environmental degradation.

In this context, 'lack of full scientific certainty' (or 'scientific uncertainty') signals that, in the expert scientific community, there are different views about whether some action or practice, X, causes harmful outcome, Z, or about the likelihood of Z eventuating. Clearly, regulators need to factor into their calculations the fact that such differences of expert opinion exist. The question is: in such a context, is it rational for regulators to take a precautionary approach?

In many international instruments, given scientific uncertainty, a precautionary approach has crystallised into a version of the PP. However, critics of the PP argue that it is not a rational basis for regulatory intervention. For example, Gary Marchant and Douglas Sylvester (2006: 722) have accused it of being 'an overly-simplistic and under-defined concept that seeks to circumvent the hard choices that must be faced in making any risk management decision.' In some small part, the problem is that the PP can be articulated and interpreted in many different ways, but the major objection is that a precautionary approach is apparently urged not only without taking any account of the cost of the intervention (in particular, the loss of whatever value or benefit X has) but also without it being certain that X will cause (or is already causing) Z. There is no need to rehearse this well-worn debate, but before introducing the relevance of regulatory stewardship and the protection of the global commons, one short comment is in order.

It will be appreciated that one function of the PP is to challenge regulators who routinely respond to scientific uncertainty by procrastinating. Although, as Jonathan Zittrain (2008) has argued, regulatory procrastination has served us well in the development of information technologies (where regulators did not try to anticipate and prevent the occurrence of any harmful effects), it surely does not pass muster as an across-the-board rational response. In other words, it is irrational for regulators routinely to eschew precautionary intervention until there is full scientific certainty. Conversely, it is not rational for regulators routinely to respond to scientific uncertainty by making a precautionary intervention. This is not to say that a precautionary intervention is never rationally justified, but such an intervention will only be a prudent and responsible response where it follows from an all-things-considered judgement (including taking into account the loss of benefit occasioned by the intervention).

Stewardship

Our analysis of the regulatory responsibilities (in Chapter 17) invites the thought that notwithstanding criticism of the PP, a form of precautionary reasoning might well be acceptable in defence of the commons. According to such reasoning, where regulators cannot rule out the possibility that some activity threatens the deepest infrastructure (which, on any view, is potentially 'catastrophic'), then they should certainly engage a precautionary approach. This reasoning, it should be emphasised, assumes an active employment of precaution. It is not simply that a lack of full scientific certainty is no reason (or excuse) for inaction – which puts one reason for inaction out of play but still has no tilt towards action; rather, where the harm concerns the commons, there is a need to initiate preventive and protective action.

The range of precautionary measures is quite broad. At minimum, regulators should invest some resources in understanding more about whether X is causing, or is likely to cause, Z; they should consider withdrawing any IP encouragement (notably patents) for the relevant technology; and they may in good faith apply protective measures or prohibitions even though such measures involve some sacrifice of a valued activity (actual or anticipated). It is true that, with the benefit of hindsight, it might be apparent that a precautionary sacrifice was actually unnecessary. However, the alternative is to decline to make the sacrifice even when this was necessary to defend the generic conditions. If regulators gamble with the essential infrastructure, and if they get it wrong, it is not just the particular valued activity, but all human activities, that will be affected adversely.

In Law 3.0, we should also expect there to be attention to possible technical solutions to threats to the commons (such as adverse climate change). For example, in this vein, it might be argued that, if humans will not comply with normative regulatory requirements that are designed to tackle global warming, a non-normative geo-engineering technical fix might be a legitimate way of dealing with the problem (see, e.g., Reynolds, 2011). Of course, the conditions on which technological management is licensed for the purpose of protecting the commons need to be qualified. It makes no sense to trade one catastrophe for another. So bilateral precaution needs to be applied: if interventions that are containable and reversible are available, they should be preferred. If in doubt, perhaps regulators should give rules a fair chance to work before resorting to experimental technological management.

This leaves one other point. Characteristically, infrastructures are of value because they support a range of purposes. Not all infrastructures are as deep as the commons but their value should not be discounted. Political communities might well decide that modern infrastructures, such as the Internet, are so valuable that they need to be protected by measures of technological management. Provided that these communities have a clear understanding of not only the value of such infrastructural resources but also of any downsides to technological management, then they should be entitled to regulate the resource in this way.

Chapter 19

Reinventing the Rule of Law

Technical solutions in general and technological management in particular appeal because of their promise to be more effective than rules, but the brute instrumentalism of technical measures demands that their use is conditioned by principles that give them legitimacy – otherwise there is no reason why regulatees should at least acquiesce in their use. Enter the Rule of Law operating in tandem with the scheme of regulatory responsibilities (the benchmarks of legitimacy) sketched in Chapter 17.

Even though there are many conceptions of the Rule of Law, I take it that this is an ideal that sets its face against both arbitrary governance and irresponsible citizenship. Advocates of particular conceptions of the ideal will specify their own favoured set of conditions (procedural and substantive, thin or thick) for the Rule of Law, which in turn will shape how we interpret the line between arbitrary and non-arbitrary governance as well as whether we judge citizens to be acting responsibly or irresponsibly in their response to acts of governance. Viewed in this way, the Rule of Law represents a compact between, on the one hand, lawmakers, law-enforcers, law-interpreters, and law-appliers, and on the other hand, the citizenry. The understanding is that the actions of those who are in the position of the former should always be in accordance with the authorising constitutive rules (with whatever procedural and substantive conditions are specified) and that, provided that the relevant actions are in accordance with the constitutive rules, then citizens (including lawmakers, law-enforcers, law-interpreters, and law-appliers in their capacity as citizens) should respect the legal rules and decisions so made. In this sense, no one – whether acting offline or online – is above the law, and the Rule of Law signifies that the law rules.

Applying this ideal to the acts of regulators (whether these are acts that set standards, that monitor compliance, or that take corrective steps in response to non-compliance), those acts should respect the constitutive limits and, in turn, they should be respected by regulatees provided that the constitutive rules are observed.

In principle, we might also – and, indeed, I believe that we should – apply the ideal of the Rule of Law to the use of technological measures. In particular, the fact that regulators who employ technological management resort to a non-normative instrument does not mean that the compact is no longer relevant. On the one side, it remains important that the exercise of power through technological management is properly authorised and limited, and on the other – although citizens might have less opportunity for 'non-compliance'– it is important that the constraints imposed by technological management are respected. To be sure, the context of regulation by technological management is very different to that of a normative legal environment, but the spirit and intent of the compact remains relevant.

The importance of the Rule of Law in an era of technological management should not be understated. Indeed, if we are to reinvent law for our technological times, one of the first priorities is to shake off the idea that brute force and coercive rules are the most dangerous expressions of regulatory power; the regulatory power to limit our practical options might be much less obvious but no less dangerous. Power, as Steven Lukes (2005: 1) rightly says, 'is at its most effective when least observable.'

While I cannot here specify a model Rule of Law for future communities, I suggest that the following conditions merit serious consideration.

First, for any community, it is imperative that technological management (just as with rules and standards) does not compromise the essential conditions for human social existence (the commons). The Rule of Law should open by emphasising that the protection and maintenance of the commons is always the primary responsibility of regulators.

Second, where the aspiration is not simply to be a moral community (a community committed to the primacy of moral reason) but a particular kind of moral community, then it will be a condition of the Rule of Law that the use of technological management (just as with rules and standards) should be consistent with its particular constitutive features – whether those features are, for instance, liberal or communitarian in nature, rights-based or utilitarian, and so on.

Looking ahead, one thought – a thought that has already occurred – is that a community might attach particular value to preserving both human officials (rather than machines) and rules (rather than technological measures) in the core areas of the criminal justice system. Indeed, it might be suggested that core crime should be ring-fenced against technological management. In this way, an important zone for moral development (and display of moral virtue) will be preserved and will retain some flexibility in the application of the rules.

Third, where the use of technological management is proposed as part of a risk management package, so long as the community is committed to the ideals of deliberative democracy, it will be a condition of the Rule of Law that there needs to be a transparent and inclusive public debate about the terms of the package. It will be a condition that all views should be heard with regard to whether the package amounts to both an acceptable balance of benefit and risk and whether it represents a fair distribution of such risk and benefit (including adequate compensatory provisions). Before the particular package can command respect, it needs to be somewhere on the spectrum of reasonableness. This is not to suggest that all regulatees must agree that the package is optimal; but it must at least be reasonable in the weak sense that it is not a package that is so unreasonable that no rational regulator could, in good faith, adopt it.

For example, where technologically managed places or products operate dynamically, making decisions case-by-case or situation-by-situation, then one of the outcomes of the public debate might be that the possibility of a human override is reserved. In the case of driverless cars, for instance, we might want to give agents the opportunity to take control of the vehicle in order to deal with some hard moral choice (whether of a 'trolley' or a 'tunnel' nature) or to respond to an emergency (perhaps involving a 'rescue' of some kind). Beyond this, we might want to reserve the possibility of an appeal to humans against a decision that triggers an application of technological management that forces or precludes a particular act or that excludes a particular person or class of persons. Indeed, the concern for last resort human intervention might be such a pervasive feature of the community's thinking that it is explicitly embedded as a default condition in the Rule of Law.

Similarly, there might be a condition that interventions involving technological management should be reversible – a condition that might be particularly important if measures of this kind are designed not only into products and places but also into people, as might be the case if regulators contemplate making interventions in not only the coding of product software but also the genomic coding of particular individuals.

Fourth, where following community debate or public deliberation, particular limits on the use of technological management have been agreed, those limits should be respected. Clearly, it would be an abuse of power to exceed such limits. In this sense, the use of technological management should be congruent with the particular rules agreed for its use and coherent with the community's constitutive rules.

Fifth, the community will want to be satisfied that the use of technological measures is accompanied by proper mechanisms for accountability. When there are problems, or when things go wrong, there need to be clear, accessible, and intelligible lines of accountability. It needs to be clear who is to be held to account as well as how they are to be held to account, and the accounting itself must be meaningful.

Sixth, a community might be concerned that the use of technological management will encourage some mission creep. If so, it might stipulate that the restrictive scope of measures of technological management or their forcing range should be no greater than would be the case were a rule to be used for the particular purpose. In this sense, the restrictive sweep of technological management should be, at most, coextensive with that of the equivalent (shadow) rule.

Seventh, it is implicit in Lon Fuller's (1969) well-known principles of legality that regulators should not try to trick or trap regulatees, and this is a principle that is applicable whether the instrument of regulation is the use of rules or the use of technological management. Accordingly, it should be a condition of the Rule of Law that technological management should not be used in ways that trick or trap regulatees, and that in this sense the administration of a regime of technological management should be in line with the reasonable expectations of regulatees (implying that regulatees should be put on notice that technological management is in operation). Crucially, if the default position in a technologically managed regulatory environment is that, where an act is found to be available, it should be treated as permissible, then regulatees should not be penalised for doing the act on the good faith basis that, because it is available, it is a lawful option.

Eighth, regulatees might also expect there to be a measure of public authorisation and scrutiny of the private use of technological management. Indeed, as Julie Cohen (2019: 267) puts it, it is self-evident that 'institutions for recognising and enforcing fundamental rights should work to counterbalance private economic power rather than reinforcing it. Obligations to protect fundamental rights must extend – enforceably – to private, for-profit entities if they are to be effective at all.' The point is that, even if public regulators respect the conditions set by regulatees, it will not suffice if private regulators are left free to use technological management in ways that compromise the community's moral aspirations, violate its constitutive principles, or exceed the agreed and authorised limits for its use. Accordingly, it should be a condition of the Rule of Law that the *private* use of technological management should be compatible with the general principles for its use.

Summing up, if we compare the Rule of Law so reinvented with other articulations of the idea, we will notice the following three key differences: (a) the scope of the Rule of Law includes governance by technological measures as much as by rules; (b) it applies to both public and private regulatory interventions; and (c) over and above the usual procedural conditions (which demand, among other things, that the rules should be published, general, and prospective, that their requirements should be clear, and that their administration should be as declared) (see Fuller, 1969), there are major substantive conditions, most importantly reminding regulators of their responsibilities in relation to the preservation and maintenance of the global commons.

Chapter 20

Technology and the triple licence

This is a very short chapter, but it takes a chapter to highlight the importance of one of the key threads of the discussion. This is the idea that no technological instruments should be applied for regulatory purposes unless they meet the terms of a triple licence, namely: a (global) commons licence, a community licence, and a social licence. In other words, in an age of technological management, regulatory legitimacy should hinge on the measures in question – whether employed by public or by private regulators – satisfying the terms of this triple licence.

Following the pattern of the benchmarks of legitimacy as represented by the three-tiered understanding of regulatory responsibilities, and as then embedded in an extended articulation of the Rule of Law, each part of the triple licence sets its own conditions for the use of technological measures. The first and most important element of the triple licence is that the measures in question must be compatible with respect for the preconditions for human social existence (with the preconditions that constitute the global commons). The second strand of the triple licence, the community licence, demands that, within a particular community, the application of technological measures should be compatible with the fundamental values of that community – with the values, so to speak, that give the community its distinctive identity, that make it the particular community that it is. Finally, the third element of the triple licence is the social licence, a licence that hinges on regulators reaching a reasonable accommodation of whatever plurality of views (for example, views about the importance of innovation and the balance of benefits and risks) there might be in their community (and which they identify in their consultative and deliberative processes).

Now it is sometimes said that we need to have a new social contract or compact for the application of modern technologies (see, e.g., Lucassen, Montgomery, and Parker, 2016), and the triple licence might be presented as just such a contract or compact. However, it needs to be understood that the first element of the licence is fundamental to each and every community's contract; this is a non-negotiable standard term. Once we get to the second and third elements of the licence, each community has some freedom to take its own distinctive position (compare House of Lords Select Committee on Artificial Intelligence, 2017). It is, of course, essential that the fundamental values to which a particular community commits itself, or its accommodations of plurality, are consistent with (or cohere with) the commons' conditions. Provided that this is the case, then regulatory legitimacy turns on maintaining fidelity with the community's constitutive values and taking up positions that are somewhere in the range of reasonableness. No doubt there will be many interpretive questions here, but the exercise is internal to the commitments of the particular community and its practice in accommodating competing and conflicting interests. Accordingly, in principle, the fundamental values of Community A might be quite different to Community B and again to Community C, and the practical accommodations in Community A might also be different to those reached in Community B and again in Community C. At this level, there can be a plurality of communities, each with their own distinctive community and social licences.

It follows that there might be many triple licences, displaying different cultural preferences and different community-defining aspirations, but in all places there should be no green light for technological management unless it meets the requirements of the non-negotiable commons licence.

Chapter 21

High-tech policing and crime control

Broadly speaking, the recent history of criminal justice reflects a tension between the politics of crime control ('law and order') and the civil libertarian demand for due process (Packer, 1969). While due process requirements are seen by advocates of crime control as an unwelcome 'obstacle course' standing between law enforcement officers and the conviction of the guilty, they are viewed by civil libertarians as essential protections of both innocent persons and the rights of citizens. With the rise of a Law 2.0 regulatory-instrumentalist mindset, and with the reduction of crime as the primary regulatory objective, we should expect there to be increasing pressure on the due process features of criminal justice systems, and at any rate in the United Kingdom the conventional wisdom is that these features have been weakened and watered down.

As Law 3.0 takes hold to become the dominant regulatory mentality, we can expect there to be a focus on the effective management of risk, on *ex ante* prevention, and on the use of technological measures. In this context, Amber Marks, Ben Bowling, and Colman Keenan (2017: 705) suggest that the direction of travel is towards

> an increasingly automated justice system that undercuts the safeguards of the traditional criminal justice model. This system favours efficiency and effectiveness over traditional due process safeguards and is taking on a life of its own as it becomes increasingly mediated by certain types of technology that minimize human agency.

To a considerable extent, this vision of automated justice anticipates the rapid development and deployment of smart machines, but Marks, Bowling, and Keenan also highlight the significance of Big Data, surveillance, and the 'new forensics.'

The new forensics

The new forensics, particularly the making, retention, and use of DNA profiles, has been with us for some time. In the United Kingdom, advocates of crime control saw this biotechnological breakthrough as an important tool for the police and prosecutors, and the legislative framework was duly amended to authorise very extensive taking and retention of profiles. Even when legal proceedings were dropped or suspects were acquitted, the law authorised the retention of the profiles that had been taken. As a result, a DNA database with several million profiles soon was in place, and where DNA samples were retrieved from crime scenes, the database could be interrogated as an investigative tool (so that 'reasonable suspicion' could be cast on an individual, not by independent evidence, but by a 'match'). Precisely how much contribution to crime control was (or is) made by the profiles is hard to know. However, it was clear that the traditional rights of individuals were being subordinated to the promise of the new technology; and it was just a matter of time before the compatibility of the legislative provisions with human rights was raised in the courts. Famously, in the *Case of S. and Marper v The United Kingdom* (2009), the leading case in Europe on the taking (and retention) of DNA samples and the banking of DNA profiles for criminal justice purposes, the Grand Chamber in Strasbourg held that the legal provisions were far too wide and disproportionate in their impact on privacy. To this extent at least, individual human rights prevailed over the latest technology of crime control (Brownsword and Goodwin, 2012: Ch 4).

Although DNA profiling and the 'new forensics' (including digital fingerprinting) offer important investigative resources, these technologies are still operating *after* the event, after a crime has been committed. However, the logic of Law 3.0 is to seek out technologies that are capable of being used in ways that promise to operate *before* the event, anticipating and preventing the commission of crime. This is where Big Data, facial recognition, machine learning, and artificial intelligence, operating in conjunction with human preventive agents and the required technological resources, pave the way for the automation of criminal justice.

Automated criminal justice

The automation of criminal justice does not happen overnight; it is an incremental process. One of the increments is from the use of technologies to advise and to assist the police and other criminal justice professionals to their use as decision-makers in their own right, and concomitantly another increment is from humans being in the loop to humans being out of the loop.

(i) From advice and assistance to replacement

To the extent that discretion and risk assessment are built into the administration of the criminal justice system, there is no avoiding them. However, with the assistance of smart tools, the exercise of human discretion might be 'regularised,' it might be applied more consistently, it might be abused less frequently, and overall it might be rendered more acceptable. That said, how might the use of AI in sentencing fare if challenged directly on due process grounds?

Precisely such a challenge was mounted in the case of *State of Wisconsin v Loomis* (2016), where the defendant denied involvement in a drive-by shooting but pleaded guilty to a couple of less serious charges. The circuit court, having accepted the plea, ordered a Presentence Investigation Report (PSI) to which a COMPAS risk assessment was attached. That assessment showed the defendant as presenting a high risk of recidivism, and the court duly relied on the assessment along with other sentencing considerations to rule out probation. In response to the defendant's appeal on due process grounds, the Wisconsin Court of Appeals certified a number of questions for the Wisconsin Supreme Court which ruled against the defendant in the following terms (at 753–754):

8 Ultimately, we conclude that if used properly, observing the limitations and cautions set forth herein, a circuit court's consideration of a COMPAS risk assessment at sentencing does not violate a defendant's right to due process.

9 We determine that because the circuit court explained that its consideration of the COMPAS risk scores was supported by other independent factors, its use was not determinative in deciding whether Loomis could be supervised safely and effectively in the community. Therefore, the circuit court did not erroneously exercise its discretion.

The relevant 'limitations and cautions' were set out by the Court as follows (at 769):

98 [A] sentencing court may consider a COMPAS risk assessment at sentencing subject to the following limitations. As recognized by the Department of Corrections, the PSI instructs that risk scores may not be used: (1) to determine whether an offender is incarcerated; or (2) to determine the severity of the sentence. Additionally, risk scores may not be used as the determinative factor in deciding whether an offender can be supervised safely and effectively in the community.

99 Importantly, a circuit court must explain the factors in addition to a COMPAS risk assessment that independently support the sentence imposed. A COMPAS risk assessment is only one of many factors that may be considered and weighed at sentencing.

100 Any Presentence Investigation Report ('PSI') containing a COMPAS risk assessment filed with the court must contain a written advisement listing the limitations. Additionally, this written advisement should inform sentencing courts of the following cautions as discussed throughout this opinion:

- The proprietary nature of COMPAS has been invoked to prevent disclosure of information relating to how factors are weighed or how risk scores are determined.
- Because COMPAS risk assessment scores are based on group data, they are able to identify groups of high-risk offenders – not a particular high-risk individual.

- Some studies of COMPAS risk assessment scores have raised questions about whether they disproportionately classify minority offenders as having a higher risk of recidivism.
- A COMPAS risk assessment compares defendants to a national sample, but no cross-validation study for a Wisconsin population has yet been completed. Risk assessment tools must be constantly monitored and re-normed for accuracy due to changing populations and subpopulations.
- COMPAS was not developed for use at sentencing, but was intended for use by the Department of Corrections in making determinations regarding treatment, supervision, and parole.

101 It is important to note that these are the cautions that have been identified in the present moment. For example, if a cross-validation study for a Wisconsin population is conducted, then flexibility is needed to remove this caution or explain the results of the cross-validation study. Similarly, this advisement should be regularly updated as other cautions become more or less relevant as additional data becomes available.

Although this might be seen as the thin end of the AI wedge (compare Chapter 16), the limitations and cautions enumerated by the court reflect some important pressure points concerning the acceptability of the use of tools such as COMPAS. Moreover, in a concurring opinion, the Chief Justice emphasises that, although the Court's holding 'permits a sentencing court to *consider* COMPAS, we do not conclude that a sentencing court may *rely* on COMPAS for the sentence it imposes' (772). The legitimate function of COMPAS, in other words, is to assist judges, not to replace them.

Nevertheless, if smart machines are perceived to outperform human decision-makers and risk-assessors (relative to criteria of accuracy, consistency, fairness, and so on), then it is likely to be just a matter of time before the technologies go beyond advising and assisting humans.

(ii)   From being in the loop to being out of the loop

Article 22 of the GDPR endeavours to keep humans in the loop where automated decision-making threatens significant human interests. However, it is the Law Enforcement Directive (Directive 2016/6801) that makes specific provision for automated processing in the criminal justice system. According to Article 11(1) of the Directive:

Member States shall provide for a decision based solely on automated processing, including profiling, which produces an adverse legal effect concerning the data subject or significantly affects him or her, to be prohibited unless authorised by Union or Member State law to which the controller is subject and which provides appropriate safeguards for the rights and freedoms of the data subject, at least the right to obtain human intervention on the part of the controller.

In order to claim the protection of these provisions, the data subject must show (a) that there has been a *decision* based *solely* on automated processing (b) which has produced adverse legal effects or (c) which has *significantly* affected him or her. Lawyers will detect several nice points of interpretation here.

In particular, how should we read the threshold condition of a decision that is based 'solely' on automated processing? For example, would we say that the processing of offences by cameras/videos worn by policemen is solely automated? How relevant is it that the camera has to be switched on by, as well as being worn by, a human police officer? Is this materially different to the automated processing of road traffic offences by roadside cameras? Again, while it is easy enough to think of examples of adverse legal effects (such as a denial of bail or a denial of parole), what might count as 'significant' effects? There is also the question of what counts as a 'decision.' In the smart environments of the future, where the controlling technologies are connected, embedded, and integrated, what does it takes for a 'decision' to stand out from the background humming of the machines?

Even if these interpretive issues can be satisfactorily resolved, how reassuring are the safeguards? How effective is the possibility of recourse to human intervention likely to be? In an age when AI and automated decisions outperform humans, how realistic, reasonable, or rational is it for humans, having reconsidered the matter, to override the automated decision (see Hin-Yan Liu, 2018)? The point is that if AI is 'trusted,' then the possibility of bringing humans back into the loop might be little more than an empty gesture. On the one hand, as with many ostensibly remedial pathways, the gradient is simply too steep; even for those prospective complainants who know about the availability of a remedy, the cost and complexity of pursuing a complaint is just too great. On the other hand, the humans who are brought back into the loop might be reluctant to gainsay the automated decision – in which case, this will further disincentivise individual complainants. If humans are to be brought back into the loop, and if smart machines are to be effectively monitored, it is probably not at the behest of individual complainants. Rather, it will be left to regulatory bodies to undertake *ex ante* licensing of AI and *ex post* audit of its performance.

Does it end well?

According to Stephen Hawking (2018: 188), 'the advent of super-intelligent AI would be either the best or the worst thing ever to happen to humanity.' At best, smart machines, smart policing, and smart cities might signal the end of crime, but at worst we can imagine various dystopian futures where the existential and agential threats presented by AI have been realised. Given, in James Bridle's words (2018: 2), that our technologies are complicit in an out-of-control economic system that immiserates many and continues to widen the gap between rich and poor; the collapse of political and societal consensus across the globe resulting in increasing nationalisms, social divisions, ethnic conflicts and shadow wars; and a warming climate, which existentially threatens us all,

then some might think that automating justice might not be the smartest way of trying to achieve the end of crime.

Chapter 22

The renewal of coherentism

In the bigger picture of regulatory responsibilities (see Chapter 17), which mind-set should regulators engage? Should regulators reason like coherentists, focusing on the integrity of doctrine and applying traditional principles? Or should they reason in a regulatory-instrumentalist policy-focused way, or in a technocratic way looking for technical solutions? How are the regulatory responsibilities to be most effectively discharged?

Given that the paramount responsibility is to ensure that no harm is done to the commons, we might be concerned that, if regulators think in a traditional coherentist way, they might fail to take the necessary protective steps – steps that might involve new rules, or the use of measures of technological management, or both. This suggests that a regulatory-instrumentalist approach is a better default, but it is only so if regulators are focused on the relevant risks – namely, the risks presented by technological development to the commons' conditions. Moreover, we might want to add that regulatory-instrumentalism with this particular risk focus is only a better default if it is applied with a suitably precautionary mentality (compare my remarks in Chapter 18). Regulators need to understand that compromising the commons is always the worst-case scenario. Alongside such a default, a technocratic approach might well be appropriate. For example, as we have said, if we believe that a rule-based approach cannot protect the planetary boundaries, then a geo-engineering approach might be the answer.

Accordingly, an essential element in living with Law 3.0 is the articulation of a 'new coherentism,' reminding regulators of two things: first, that their most urgent regulatory focus should be on the commons' conditions, and second, that whatever their interventions, and particularly where they take a technocratic approach, their acts must always be compatible with the preservation of the commons.

In the future, the courts – albeit the locus for traditional coherentist thinking – will have a continuing role to play in bringing new coherentism to bear on the use of technological measures (with a central question being whether a particular use meets the terms of the triple licence). In other words, it will be for the courts to review the legality of any measure that is challenged relative to the authorising and constitutive rules, and above all, to check that particular instances of technological management are consistent with the commons-protecting ideals that are inscribed in the Rule of Law.

With a new coherentist mindset, it is not a matter of checking for internal doctrinal consistency, nor checking that a measure is fit for its particular regulatory purpose. Rather, a renewed ideal of coherence should start with the paramount responsibility of regulators, namely, the protection and preservation of the commons. All regulatory interventions should cohere with that responsibility. This means that the conditions for both human existence and the context for flourishing agency should be respected. In line with such thinking, in 2017, when researchers met at Asilomar in California to develop a set of precautionary guidelines for the use of AI, it was agreed (in Principle 21) that 'risks posed by AI systems, especially catastrophic or existential risks, must be subject to planning and mitigation efforts commensurate with their expected impact' (Asilomar Conference, 2017).

Moreover, as we have emphasised, if the commons is to be respected, technological management should not be employed in ways that compromise the context for agency and moral community. Consider, for example, the much debated and protean concept of privacy. A popular view is that respect for privacy should be applied in a 'contextual' way. However, there is Context and there are contexts. There is Context (in the sense of the commons) and then there are many contexts that rely on the integrity of the commons. So if it is judged that privacy reaches through to the interests that agents necessarily have in the commons' conditions, particularly in the conditions for self-development and agency, it is neither rational nor reasonable for agents, individually or collectively, to authorise acts that compromise these conditions (unless they do so in order to protect some more important condition of the commons) (compare Brincker, 2017: 64). That is, as Bert-Jaap Koops (2018: 621) has so clearly expressed it, privacy has an 'infrastructural character,' 'having privacy spaces is an important presupposition for autonomy [and] self-development'; without such spaces, there is no opportunity to be oneself. On this reading, privacy is not so much a matter of protecting goods (informational or spatial) in which one has a personal interest but protecting infrastructural goods in which there is either a common interest (engaging first-tier responsibilities) or a distinctive community interest (engaging second-tier responsibilities).

On the other hand, if privacy (and, likewise, data protection) is judged to be simply a legitimate informational interest that has to be weighed in an all-things-considered balance of interests, then we should recognise that what each community will identify as a privacy interest and as an acceptable balance of interests might well change over time. To this extent, our reasonable expectations of privacy might be both 'contextual' and contingent on social practices. That said, a community might wish to define itself by giving privacy an elevated status (as a right or a fundamental right) which regulators will then need to respect as an overriding interest. However, no community can rationally define itself in ways that are incompatible with the common interest in the essential infrastructural conditions.

Next, measures of technological management should cohere with the particular constitutive values of the community – such as respect for human rights and human dignity, the way that non-human agents are to be treated, and so on – and its particular articulation of the Rule of Law. Expressing a concern that AI should be 'trustworthy,' there has been a proliferation of codes and guidelines. For example, the aforementioned Asilomar researchers agreed that 'AI systems should be designed and operated so as to be compatible with ideals of human dignity, rights, freedoms, and cultural diversity' (Principle 11), and in 2019 the EC high-level expert group on AI, the OECD, and the G20 each endorsed a 'human-centric' approach to AI. No doubt, the courts will face many challenges in developing a coherent account of these principles, but their principal task will be to ensure that particular instances of technological management cohere with, rather than abuse, these or similar principles.

Coherence, as in Law 1.0, might be an ideal that is dear to the hearts of private lawyers. However, in an era of Law 3.0, only once coherence is brought into the body of public law can we see its full regulatory significance. Unlike in Law 1.0, where coherence is about doctrinal consistency, and Law 2.0, where coherence is about the complementarity of various measures, in Law 3.0 coherence is about compatibility with the benchmarks of legitimacy. Regulation, whether normative or non-normative, will lack coherence if the procedures or purposes that accompany it are out of line with the authorising or constitutive rules that take us back to the Rule of Law itself, and regulation will be fundamentally incoherent if it is out of line with the stewardship responsibility for maintaining the commons. In short, we can continue to treat coherence as an ideal that checks backwards, sideways, and upwards, but the reimagination of this ideal necessitates its engagement with the triple licence for technologies, with the full range of regulatory responsibilities and with the full repertoire of regulatory instruments.

Chapter 23

Redesigning the institutional framework I

National institutions

If we are to be properly geared for the discharge of regulatory responsibilities, this might call for some redesigning of the institutions on which we rely both nationally and internationally. While we can expect national regulators to deal with the routine balancing of interests within their communities as well as respecting the distinctive values of their particular community, the stewardship of the commons seems to call for international oversight.

In this chapter, we can start with some remarks about the arrangements nationally for engaging with emerging technologies and then, in the next chapter, we can turn to the possible international regulation of the commons.

In the United Kingdom (and, I suspect, in many other nation states), there are two contrasting features in the institutional arrangements that we have for engaging with and regulating new technologies. On the one hand, there is no standard operating procedure for undertaking the initial review of such technologies, and on the other hand, the Rule of Law in conjunction with democracy dictates that the courts should settle disputes in accordance with established legal principles and that it is for the Legislature and the Executive to formulate and agree on public policies, plans, and priorities. In other words, while there is no expectation about who will undertake the initial review or how that review will be approached, we have very definite expectations about the role and reasoning of judges and advocates in the courts (where Law 1.0 applies) and similarly about the policy-making members of the Legislature and Executive (where Law 2.0 applies). The question is: where in this institutional design do we find the responsibility for stewardship of the commons and for the community's distinctive values?

To start with the initial engagement with, and review of, an emerging technology, it seems to be largely a matter of happenstance as to who addresses the issue and how it is addressed. For example, in the late 1970s, when techniques for assisted conception were being developed and applied, but also being seriously questioned, the response of the UK government was to set up a Committee of Inquiry chaired by Mary Warnock. In 1984, the Committee's report (the Warnock Report) was published. However, it was not until 1990, and after much debate in Parliament, that the framework legislation, the Human Fertilisation and Embryology Act 1990, was enacted. This process, taking the best part of a decade, is regularly held up as an example of best practice when dealing with emerging technologies. Nevertheless, this methodology is not in any sense the standard operating procedure for engaging with new technologies – indeed, there is no such procedure.

The fact of the matter is that legal and regulatory responses to emerging technologies vary from one technology to another, from one legal system to another, and from one time to another. Sometimes, there is extensive public engagement, sometimes not. On occasion, special commissions (such as the now defunct Human Genetics Commission in the UK) have been set up with a dedicated oversight remit, and there have been examples of standing technology foresight commissions (such as the US Office of Technology Assessment). Often, however, there is nothing of this kind. Most importantly, questions about new technologies sometimes surface, first, in litigation (leaving it to the courts, applying the logic of Law 1.0 to determine how to respond) and, at other times, they are presented to the legislature (as was the case with assisted conception).

With regard to the question of which regulatory body engages with new technologies and how, there can of course be some local agency features that shape the answers. Where, as in the United States, there is a particular regulatory array with each agency having its own remit, a new technology might be considered in just one lead agency or it might be assessed in several agencies. Once again, there is a degree of happenstance about this. Nevertheless, in a preliminary way, we can make three general points.

First, if the question (such as that posed by a compensatory claim made by a claimant who alleges harm caused by a new technology) is put to the courts, their responsibility for the integrity of the law will push them towards a coherentist, Law 1.0, assessment. Typically, courts are neither sufficiently resourced nor mandated to undertake a risk assessment let alone adopt a risk-management strategy.

Second, if the question finds its way into the legislative arena, it is much more likely that politicians will engage with it in the regulatory-instrumentalist way that is characteristic of Law 2.0, and once the possibility of technological measures gets onto the radar, it is much more likely that (as with institutions in the EU) we will see a more technocratic, Law 3.0, mindset.

Third, if leaving so much to chance seems unsatisfactory, then it is arguable that there needs to be a body that is charged with undertaking the preliminary engagement with new technologies. The remit and challenge for such a body would be to ensure that there is no harm to the commons, to try to channel such technologies to our most urgent needs (relative to the commons), and to help each community to address the question of the kind of society that it distinctively wants to be – doing all that, moreover, in a context of rapid social and technological change. As Wendell Wallach (2015: 10) rightly insists:

> Bowing to political and economic imperatives is not sufficient. Nor is it acceptable to defer to the mechanistic unfolding of technological possibilities. In a democratic society, we – the public – should give approval to the

futures being created. At this critical juncture in history, an informed conversation must take place before we can properly give our assent or dissent.

Granted, the notion that we can build agencies that are fit for such purposes might be an impossible dream. Nevertheless, all impossible dreamers must agree that this is the right time to set up a suitably constituted body – possibly along the lines of the Centre for Data Ethics and Innovation (to inform the public and to set standards for the ethical use of AI and data) – that would underline our responsibilities for the commons as well as facilitate the development of each community's regulatory and social licence for these technologies.

To return to our earlier discussion of autonomous vehicles, it makes little sense to try, in the fashion of Law 1.0, to apply the principles for judging the negligence of human drivers where there is no human in control and where the nature of the technology militates against simple causal accounts when things 'go wrong.' Yet if these questions are taken up in the courts, we must expect that judges (reasoning like coherentists) will try to apply notions of a reasonable standard of care, proximate cause, and so on, to determine responsibility for very complex technological failures. Indeed, when Joshua Brown was killed while driving his Tesla S car in autopilot mode, Tesla (presumably anticipating litigation or a discourse of fault and responsibility) were quick to highlight the safety record of their cars, to suggest that drivers of their cars needed to remain alert, and to deny that they themselves were careless in any way. By contrast, if regulators in a legislative setting approach the question of liability and compensation with a risk-management mindset, they will not need to chase after questions of fault – or, at any rate, as in the UK Automated and Electric Vehicles Act 2018, insurance and compensation will come first with insurers or owners of automated vehicles then able to pursue existing (fault-based) common law claims. Reasoning in a Law 2.0 way, the challenge will be to articulate the most acceptable (and financially workable) compensatory arrangements that accommodate the interest in transport innovation with the interest in the safety of passengers and pedestrians. That said, if regulators are to act in a way that is socially licensed, they should take a view only after an independent emerging technologies body (of the kind that we do not, but surely should, have) has informed and stimulated public debate.

Chapter 24

Redesigning the institutional framework II

International institutions

The global commons is not confined to particular nation states. The conditions for human social existence on planet Earth are relevant to all nation states and can be impacted by each nation state's activities. Accordingly, if the essential infrastructure for human social existence is to be secured, this implies that international law should recognise the importance of the commons and that it should be supported by a considerable degree of international coordination and shared responsibility.

On paper, there are some positive indications in international law – for example, in the cosmopolitan idea of *jus cogens* (the idea of acts that are categorically wrong in all places) and crimes against humanity. Moreover, when the United Nations spells out the basic responsibilities of states, it can do so in terms that are redolent of the commons' conditions. For instance, Article 18(b) of the UN Global Compact on Safe, Orderly and Regular Migration, which was adopted in December 2018, provides that states should invest in programmes for

> poverty eradication, food security, health and sanitation, education, inclusive economic growth, infrastructure, urban and rural development, employment creation, decent work, gender equality, and empowerment of women and girls, resilience and disaster and risk reduction, climate change mitigation and adaptation, addressing the socioeconomic effects of all forms of violence, non-discrimination, rule of law and good governance, access to justice and protection of human rights, as well as creating and maintaining peaceful and inclusive societies with effective, accountable and transparent institutions.

If we check this back against the elements of the commons' conditions that were sketched in Chapter 17, this might seem a touch overinclusive. Nevertheless, much of what Article 18(b) specifies surely aligns with the first-tier stewardship responsibilities of regulators.

In support of such provisions, there is an extensive international regulatory architecture. We might assume, therefore, that securing the commons will only require some minor adjustments or additions – rather like we might add an extension to an existing property. On the other hand, stewardship of the kind that is required calls for a distinctive and dedicated approach. It might be, therefore, that we need to have bespoke international laws and new international agencies to take this project forward. Moreover, because politics tends to operate with short-term horizons, it also implies that the regulatory stewards should have some independence from the political branch but not of course that they should be exempt from the Rule of Law's culture of accountability and justification.

That said, whatever the ideal legal provision, we have to take into account the realities of international relations.

First, while all Member States of the United Nations are formally equal, the reality is that some are more equal than others, this being exemplified by the constitution of the Security Council. Not only are the five permanent members of the Security Council amongst 'the most important actors on the world stage, given their size, their economic and financial weight, their cultural influence, and, above all, their military might' (Rosenthal, 2017: 95) but they have the power to veto (and, in practice, they do veto) decisions that the Council would otherwise make. Not surprisingly, this has led to widespread criticism of the undemocratic and unrepresentative nature of the Council and crucially, to criticism of the veto which enables the permanent members to subordinate their collective responsibilities (for the commons' conditions) to their own national priorities.

Second, as Gerry Simpson (2004) highlights, the makers and subjects of international law have different amounts of power and influence, different intentions (some are more well-intentioned than others), different levels of commitment to collective responsibilities, and different degrees of civilisation. To start with, there are both functioning states and failed states. Amongst the former, while many states are good citizens of the international order (respecting the rules of international law), there are also superpowers (who play by their own rules) and rogue states (who play by no rules). If the regulatory stewards were drawn from the good citizens, that might be fine insofar as an agency so populated would be focused on the right question and motivated by concerns for the common interest of humans. However, they would be in no position to ensure compliance with whatever precautionary standards they might propose, let alone be mandated to introduce measures of technological management.

A third reality is that, where the missions of international agencies include a number of objectives (such as trade, human rights, and environmental concerns), or where there is a dominant objective (such as the control of narcotics), value commitments (to human rights) will tend to be overridden ('collateralised') or even treated as irrelevant ('nullified'). Now, while it is one thing for the international community to unite around what it takes to be its shared prudential interests and, in so doing, to give less weight to its interest in certain aspirational values, respect for the commons' conditions should never be collateralised or nullified in this way. Accordingly, to keep this imperative in focus, if the regulatory stewards are located within an international agency, their mission must be limited to the protection of the commons, and acceptable collateralisation or nullification must be limited to non-commons matters. Even then, there would be no guarantee that the stewards would be immunised against the usual risks of regulatory capture and corruption (see Chapter 5). In short, unless the culture of international relations is supportive of the stewards, even the ideal regulatory design is likely to fail.

If the common interest is to be pursued, this is a battle for hearts and minds. As Neil Walker (2015: 199) has remarked in relation to global law, our future prospects depend on 'our ability to persuade ourselves and each other of what we hold in common and of the value of holding that in common.' An international agency with a mission to preserve the commons might make some progress in extending the pool of good citizens, but to have any chance of success (a) there needs to be a professional legal and technical/scientific community that has a clear vision of what new coherentism demands and (b) all nation states need to be on board.

Part four

Learning the law

Chapter 25

Rethinking legal education

If we are to get to grips with Law 3.0, we need to radically rethink our approach to legal education. Law 1.0 is fine for those who want to do Law 1.0-type things. However, Law 3.0 is where we are, and at a minimum, students should understand that Law 3.0 represents the bigger picture for their studies, so that when they engage in Law 1.0 conversations, they appreciate exactly where they are relative to coexistent Law 2.0 and Law 3.0 conversations. However, the curriculum might be more fundamentally revised so that, in all subjects, the principal agenda is set by Law 3.0.

In this chapter, I will sketch how, given a Law 3.0 perspective, we might approach the teaching of that most traditional of law subjects, the law of contract. There are, I suggest, three key questions to be included in the curriculum. First, how does the law of contract fit in to the wider regulatory environment for transactions? Second, as new transactional technologies become available and are taken up, should we try, like 'coherentists,' to fit these developments into the existing body of doctrine or should we think about such matters in a more 'regulatory-instrumental' way? Third, what should we make of the possibility of regulatory restrictions or requirements being, so to speak, 'designed into' the emerging technological platforms or infrastructures for contracts? In other words, what should we make of the 'technological management' of transactions?

Contract law and the regulatory environment for transactions

Our first question asks how the law of contract fits into 'the regulatory environment for transactions.' However, this is not one of the concepts that is explained in standard texts on the law of contract. Accordingly, I start by outlining the general idea of the regulatory environment for transactions as a way of framing our teaching of the law of contract and then I sketch three particular conceptions or specifications of the idea.

(i) The general idea of the regulatory environment for transactions

When, as an undergraduate in the mid-1960s, I was introduced to the law of contract, it was in the context of an embryonic modern consumer marketplace fuelled by consumer credit. In that context, the law of contract, with its commitment to freedom and sanctity of contract, found itself in something of a crisis. Evidently, the law licensed the use of standard form terms and conditions (which consumers did not read, would not understand, and played no part in negotiating) and, in particular, it licensed suppliers (of cars, refrigerators, televisions, and the like) to rely on standard term exemption clauses that put the risks of fitness, quality, and safety on the purchaser. In that context, it was said, quite rightly, that some changes to the law needed to be made, and in due course legislative changes were made. It was not said – or, at any rate, I do not recall anyone saying it – that the regulatory environment for transactions (or for consumer transactions) was not fit for purpose.

These days, I would expect the conversation to be rather different. Today, if the law licensed a dealer to leave a so-called 'car' outside the purchaser's premises when the vehicle was a shell, incapable of self-propulsion, and with various parts of the engine broken and burnt (compare the facts in *Karsales (Harrow) v Wallis* (1956)), and if this were anything like a common occurrence, then it surely would not be long before someone would suggest that the regulatory environment, like the car, was broken. After all, in the wake of one crisis, catastrophe, or scandal after another, we find some part of the blame being apportioned to a regulatory environment that has proved to be unfit for purpose. Nowadays, if we detect a crisis in the consumer marketplace, we will know that an effective response to the problem is likely to involve more than a tweak to the law of contract (or even a bespoke piece of legislation such as the Consumer Rights Act 2015). To get the regulatory environment right, it might also be necessary to make reforms in competition law, in credit law, and criminal law as well as changes in the technical standards for consumer goods and services. It might also be necessary to make changes to the regulatory agencies and, crucially, to take steps to change the business culture of those who supply goods and services in the consumer market.

To pick up the autobiographical thread, soon after I started teaching the law of contract, I became aware of the seminal work of Stewart Macaulay (1963) and Ian Macneil (1980). At this stage, 'law in context' was barely off the drawing board and it was not obvious how these commentaries might be fitted into my lecture narrative that started with offer and acceptance and ended with remedies. Even if there had been slots for 'context' in that narrative, there would have been a temptation for students to interpret whatever contextual content was supplied as marginal to the main doctrinal story. So how do insights about the practice of business contractors or about the web of social relations in and around transactions become an integral and equal part of the story?

Clearly, so long as the main story that we tell as teachers of the law of contract is the traditional Law 1.0 doctrinal story, there is a problem. Rather than starting with some segment of the law of contract, I suggest that the story needs to start with the idea that the field of interest is transactions (whether deep in the business world or in the consumer marketplace or in the emerging peer-to-peer shared economy) and that the particular focus is on the regulatory environment for transactions, that environment being understood in a broad way. However, we should not assume that the signals given by the law of contract necessarily are dominant in any particular transactional setting. There is a lot of noise in and around trans-actions, and in some environments, such as those studied by Macaulay, the signals given by the law of contract might be very weak indeed. Similarly, we know that in our everyday experience as consumers, whether offline or online, the legal signals are often much less prominent than the suppliers' particular custom and practice, their concern for their trading reputation, and the side-arrangements that have been made for guarantees and warranties. We find ourselves, as Macneil points out, in a web of social relations.

So the first lesson in contract law does not start within the law of contract itself (whether with offer and acceptance or with remedies) but with transactions and the regulatory environments in which transactors operate. The law of contract sends signals to transactors but it is just one element in the regulatory environ-ment. If we start here, the question then is just how wide we should go in drawing the boundaries of this environment.

(ii) Three conceptions of the regulatory environment for transactions

One of the challenges for any contextual approach to teaching law is to determine the scope of the context. As the context broadens, the particular doctrinal compo-nent that is ostensibly at the core of the course becomes less and less significant. If we are to start teaching the law of contract by placing it in the larger context of the regulatory environment for transactions, we have to meet this challenge. In what follows, I sketch three possible conceptions, or specifications, of this contextual idea, starting narrow and becoming broader.

A narrow (legal) specification

We might restrict the regulatory environment for transactions to the law of con-tract together with whatever other bodies of law provide signals to transactors. For example, while the law of contract provides some regulatory control for fraud and coercion, it might be argued that the primary regulatory burden is carried by the criminal law. Fraudsters, it is fair to say, are not likely to be much discouraged by a concern that their fraud, if detected, will give their co-transactors grounds for relief. Indeed, when eBay looked to the law to reinforce its reputational system, it was to the agencies of the criminal law, not to the law of contract, that it turned. Similarly, the regulatory environment for platform services in the shared economy will draw on legal resources that go beyond the law of contract.

One thing that the law of contract goes out of its way *not* to regulate is the price. A peppercorn rent is fine, but so, too, is a price that is excessive (unless it is treated as unconscionable). The price is for the parties to agree. However, when suppli-ers know so much about their customers, it becomes possible for them to engage in so-called 'dynamic pricing,' changing the price from one customer to another and even changing the price minute-by-minute in relation to the same customer. If we judge that this is unfair and an abuse of power, how is it to be regulated? The law of contract is not equipped for this task, and according to Ariel Ezrachi and Maurice Stucke (2016), neither is competition law. If this is correct, it might be the case that we should conclude that the regulatory environment is not fit for purpose and that it needs some attention. In any event, the key point is that, if there is a problem about dynamic pricing, we need to tackle it by engaging with the full sweep of the law that constitutes (on the narrow specification) the regulatory environment for transactions.

A broader (normative) specification

If we are prepared to step beyond the positive law, to recognise not just the de jure but also the de facto regulatory environment for transactions, then we can locate the law of contract in a signalling environment that includes just the kind of norms that Macaulay and Macneil identify in their work. From this perspective, we see that transactions, like interactions, take place in an ocean of normativity and that the legal norms are just one island. Here, the work of Macaulay and Macneil is part of a much larger literature that draws attention to the self-regulatory practices of groups and communities, representing so to speak the 'living law' for members of those groups and communities.

The juxtaposition of the official 'top-down' law with the unofficial 'bottom-up' law of self-regulating groups and communities raises a host of interesting issues. Where top-down law is largely viewed as being an option for transactors (as is the case with much of the law of contract), the fact that the option is not taken up is not likely to be viewed as problematic. However, where top-down regulators want transactors to adopt the official rules or, say, a state-backed payment mode or transactional technology, then resistance in some parts of the business community will lead to some turbulence in the regulatory environment. Moreover, those who sponsor new transactional technologies might find that what look like benefits on paper simply do not translate well into actual business practice.

A radical (normative and non-normative) specification

The two answers proffered so far have in common that they take the regulatory environment to be constituted by norms. The signals are normative, and they direct regulatees to what is permissible, what is required, and what is prohibited. For transactors who want to know what they ought to do in negotiating or performing a deal, or what they might reasonably expect in the event of a dispute, the regulatory environment gives some answers. We might, however, conceive of the regulatory environment for transactions in a way that reaches beyond norms to features of the technological infrastructure or platforms on which parties deal. If we do this, we allow that some elements of the regulatory environment are not normative in the sense that we think is the case with the rules of law, ethics, codes of practice, standards, and so on.

In what sense then might the signals given by particular technological features be non-normative? What the features in question signal is that a certain act in a transactional setting is or is not possible; the signals to transactors relate not to what they ought or ought not to do but to what they can and cannot do. For example, at an automated car park of the kind that English contract lawyers first debated in *Thornton v Shoe Lane Parking Ltd* (1971), the lights at the entrance to the car park might have signalled to motorists whether or not they were permitted to enter, but the architecture of the car park and the presence of the barrier at the entrance were non-normative signals of what was physically possible. Similarly, in modern online environments, unless users click to agree to the terms and conditions for accessing a website, in many cases it will not be possible to access the site. The technology is set up in a way that regulates the practical options that are actually available to users. It is not merely a matter of being normatively required to agree to certain terms and conditions, the technology ensures that without the required 'agreement' it is simply not possible to proceed.

Recalling the discussion in Chapter 15, one of the virtues of this radical specification is that it facilitates a more sophisticated analysis of the relationship between law and liberty, taking into consideration both the normative and the practical optionality of doing some act x. We can then ask whether according to some given set of laws or rules there is a 'normative liberty' to do x (whether doing x is permitted, is treated as optional) as well as whether there is a 'practical liberty' to do x in the sense that, irrespective of the rules, doing x is a real possibility, a real option.

If, in the future, we can expect the making, the performance, and the enforcement of transactions to be increasingly automated (employing a suite of emerging technologies that will support 'smart' contracting), and if the design of these technologies is such as to limit the practical options (the practical liberties) available to the parties (even to remove humans from the transactional loop), then the regulatory work is not being undertaken by the rules of the law of contract, indeed not by rules of any kind. In this future world of transactions, the regulatory environment is radically different and we might want to teach students about the law of contract in a way that highlights inter alia the relationship between normative and practical liberty by explicitly framing the law in this increasingly technological context (compare the discussion in Chapter 10).

Should we think like 'coherentists' or 'regulatory-instrumentalists'?

Regardless of how we answer the first of our three key questions (in fact, even if we reject the idea of placing the law of contract in a larger regulatory context), the emergence of new technologies can provoke questions about how the law, or how regulators, should respond.

(i) Coherentism as the default

A concern with formal coherence runs through much of the law of contract (particularly where common law and equitable doctrines are juxtaposed) as well as through critical commentaries on the state of the law. To take just one example, in *Stena Line v Merchant Navy Ratings Pension Fund Trustees Limited* (2011), Arden LJ made a significant intervention in the development of the modern jurisprudence on implied terms when she emphasised the contribution to the coherence of the law made by Lord Hoffmann's speech in *Belize Telecom* (2009):

> In *Belize*, the Privy Council analysed the case law on the implication of terms and decided that the implication of terms is, in essence, an exercise in interpretation. This development promotes the internal coherence of the law by emphasising the role played by the principles of interpretation not only in the context of the interpretation of documents *simpliciter* but also in the field of the implication of terms. Those principles are the unifying factor. The internal coherence of the law is important because it enables the courts to identify the aims and values that underpin the law and to pursue those values and aims so as to achieve consistency in the structure of the law.
>
> (para 36)

However, in a trio of more recent Supreme Court decisions – *Marks and Spencer plc v BNP Paribas Services Trust Company (Jersey) Limited* (2015) (on implied terms), and *Arnold v Britton* (2015) and *Wood v Capita Insurance Services Ltd* (2017) (both on interpretation) – we find a reaction against expansive implication and interpretation of terms, particularly in carefully drafted commercial contracts. This leaves coherentists with many questions. For some, the question is how to square contextualist with non-contextualist approaches; for others, the question is why the contextualist approach (to the extent that it has primacy) should be limited to interpretation and implication – for example, why not also determine the reasonableness of terms in a contextualist way?

By contrast, we can detect a regulatory-instrumentalist approach underlying such legislative interventions as the Arbitration Act 1979, which is conspicuously designed to serve larger economic purposes, and even in the courts where we might expect there to be less confidence in shaping the economy or taking on larger policy questions, we can detect an implicit instrumentalism in the importance attached to English commercial contract law continuing to be competitive as the choice of law in international trading communities. Nevertheless, the characteristic default of contract lawyers is to coherentism, to Law 1.0.

(ii) The significance of the distinction relative to new transactional technologies

Although coherentism centres on the internal consistency of doctrine, as we have already highlighted, it has an extended manifestation in a tendency to apply existing legal frameworks to new technological innovations that bear on transactions, or to try to accommodate novel forms of contracting within the existing categories. We need only recall *The Eurymedon* (1975) and Lord Wilberforce's much-cited catalogue of the heroic efforts made by the courts – confronted by modern forms of transport, various kinds of automation, and novel business practices – to force 'the facts to fit uneasily into the marked slots of offer, acceptance and consideration' (at 167) or whatever other traditional categories of the law of contract might be applicable.

Consider, for example, the view that online shopping sites are functionally, contextually, and normatively equivalent to offline shopping environments. Might it not be argued that it is the differences rather than the similarities that now need to be accentuated? In particular, as Ryan Calo (2014: 1002) emphasises, purchasers in online environments are technologically 'mediated consumers,' approaching 'the marketplace through technology designed by someone else.' Thanks to this technology, prices can be changed minute-by-minute and customer by customer; and the particular vulnerabilities of consumers can be identified and exploited. Even if there is nothing new in shopping environments being designed to influence purchasing decisions, in online environments this art is taken to a whole new level of technological sophistication. If there is to be a legal correction for any unfairness arising from these vulnerabilities, the question is whether the better approach is to employ a coherentist Law 1.0 doctrinal tweak or to make a bespoke Law 2.0 regulatory intervention that seeks out a more acceptable balance of interests in the online consumer marketplace.

In contrast with any lingering coherentism, we have earlier noted the conspicuously regulatory-instrumentalist approach of the European Commission in its push towards a single, and now digital, market. Prompted by such an approach, what we need to be asking is whether, by forcing the new technological facts to fit into the doctrinal slots that we have in the law of contract, we are asking the right questions. Recall, again, the conservative approach at BookWorld. Even if we can find a slot that seems to fit, perhaps we should be focusing instead on the nature of the transactional risk or problem and thinking more purposively about how to fix it. However, once we start thinking in such risk-regulatory terms, and where new technologies offer new instruments for assessing and managing risk, we have taken a significant step towards a radically different approach, a Law 3.0 approach, to the regulation of transactions.

What should we make of technical solutions and the technological management of transactions?

If we accept the radical understanding of the regulatory environment, we might not only note that some of the regulatory burden is now borne by technological design but also ask whether it is desirable to shift the burden from rules (such as those of the law of contract) in this way. In this part of the chapter, I will consider how a technocratic regulatory approach might be applied to transactions and what we might then make of this.

(i)   How would technological management be applied to transactions?

According to some commentators, contract lawyers should start to imagine a world of automated transactions, where commerce is, so to speak, a conversation conducted by machines. Instead of H2H, whether B2B or B2C contracts, we have transactions that are M2M. In line with this vision, Michal Gal and Niva Elkin-Koren (2017: 309–310) foresee a world in which

> [y]our automated car makes independent decisions on where to purchase fuel, when to drive itself to a service station, from which garage to order a spare part, or whether to rent itself out to other passengers, all without even once consulting with you.

In that world, humans have been taken out of the transactional loop, leaving it to the technology to make decisions that humans would otherwise be responsible for making. Possibly, humans will still be treated as being somewhere on the loop by treating the machines as agents transacting on behalf of human principals, neither H2H nor M2M but, as it were, (H)M2((H)M. That said, precisely how the law in general, and the law of contract in particular, will engage with that future is difficult to know.

One of the first to imagine such a future, Richard Ford (2000) foresaw that consumers would sign over their paychecks to so-called 'cyberbutlers' who would hold the funds in trust for each consumer's benefit. Then, guided by the particular consumer's profile, the cyberbutler would place appropriate orders so that, each day, the consumer would 'come home to a selection of healthy and nutritious groceries from webvan.com or a Paul Smith shirt from boo.com or the latest Chemical Brothers CD from cdnow.com' (1578). If, in place of paychecks and cyberbutlers, we now imagine virtual money, personal digital assistants, and the Internet of Things, as well as suppliers who were not dot.com failures, Ford's sense of the future is far from science fiction – indeed, it is a future that is on the near horizon in some parts of the world.

Once such supply systems are up and running, humans (whether as suppliers or as consumers) largely drop out of the picture, and technological management takes over. Now, the technology not only manages the 'ordering' and the supply of goods and services to the consumer's needs but it also manages whatever risks or problems might arise. For example, if the scheme of technological management is designed to be consumer-friendly, it will not be possible for the supplier to be credited with the payment unless the goods or services are supplied in accordance with the order and with specified standards. Conversely, if the design is supplier-friendly, the technology will ensure that consumers pay for the ordered goods and services before they are supplied. As we suggested in Chapter 10, at the root of such an arrangement, there might be a transaction, very much like a traditional contract (and to which the traditional law of contract might apply) that commits the parties to a certain protocol of technological management. Alternatively, it might be that there are background rules that prescribe the design features for systems of this kind. In this latter case, there are rules about the features of transactional technologies, but these are rules that are directed at designers and manufacturers, not at transactors. It is a new world, but before long it might be the world in which students find themselves as consumers, and if this is so, students might reasonably ask how precisely the law of contract connects to this world.

(ii) What should we make of technological management?

Suppose that digital products are supplied under contracts that require users to act in accordance with the intellectual property rights of the suppliers. Now, suppose that, instead of relying on the protection of the contractual terms and conditions, the suppliers simply code in their intellectual property rights – given the coding, it is not possible for the product to be used other than in a way that complies with the relevant rights. Thus far, we might not have serious concerns about the switch from contract to code. However, recalling our remarks in Chapter 10, if the contract or the code were to give the supplier more protection than intellectual property law recognises, this would be a problem. While a legal challenge might be mounted against the contract, the coding of products – which might not be transparent and which would be restrictive of practical liberty – might be more worrying. In both cases, though, we should insist that the actions of the suppliers should be compatible with the Rule of Law.

Suppose there is already a rule that prohibits x (such as wheeling supermarket trolleys off site and abandoning them) but because the rule is ineffective, regulators resort to technological management to eliminate the possibility of x (using GPS to redesign the trolleys so that they are immobilised once they reach their permitted limits). Even if the rule that prohibits x is superseded by technological management and 'retired,' it is not entirely redundant because it expresses the regulators' normative view (namely, that regulatees ought not to do x), and this allows for one way of testing whether a particular use of technological management satisfies the Rule of Law. Quite simply, if the *rule* to which the technological management is linked satisfies the Rule of Law, then (assuming that the technological measures are congruent with the rule, and unless there are additional requirements for the use of technological management) the particular use of technological management also satisfies the Rule of Law.

In a case where there is not already a rule that prohibits x, but where the regulator clearly believes that regulatees ought not to do x, then the use of technological management by the regulator to make it impossible to do x can be tested for Rule of Law compliance in a similar way. Here, the linkage is between the use of technological management and the rule that regulators would have put in place if they had adopted a rule that prohibits x. If such a rule would not have satisfied the Rule of Law, then the measures of technological management will also fail to do so. Conversely, if such a rule would have satisfied the Rule of Law, then technological management will also be compliant.

If we demand that measures of technological management are congruent or coextensive with rules that would be compliant with the Rule of Law, many concerns should be assuaged. However, recalling the discussion in Chapter 19, it will be a matter for debate in each community as to whether there should be additional requirements for the use of such regulatory measures – for example, requirements about transparency, reversibility, bringing humans back into the loop, and human override and oversight. There is also the question of whether, even if technological management is free of vice, there is some virtue in leaving it open to contractors to self-regulate in the customary normative way.

While it remains to be seen which particular technological infrastructures for transactions will be developed and how precisely technological management might insinuate itself into the regulation of transactions, it is already clear that emerging technologies are doubly disruptive: first, as we have said, they disrupt existing legal frameworks and concepts, alerting us to the need for new rules, and second, they disrupt the assumption that the regulatory framework is to be constituted exclusively by rules – technological assistance or management might be an option instead. Anyone who wants to wrap context around the teaching of the law of contract (or, for that matter, criminal law or torts) needs to be aware that technology has had and is continuing to have these disruptive effects.

From start to finish, a legal education in 2020 should focus on Law 3.0, on its three overlapping conversations, on its polycentric nature, and on its increasingly technological complexion. This is not legal education as we have known it, but Law 3.0 is not law as we have known it.

Chapter 26

Any questions?

Stated shortly, the key points of this introduction to Law 3.0 are the following.

First, there is Law 3.0 itself. While Law 3.0 is a certain kind of conversation, a certain way of engaging with the law, it is also a view about the scope of legal inquiry – about the field, as it were, in which lawyers should be interested. As a particular kind of conversation, Law 3.0 distinctively includes a technocratic element. Like regulatory-instrumentalist Law 2.0, but unlike coherentist Law 1.0, the Law 3.0 conversation asks whether the legal rules are fit for purpose but it also reviews in a sustained way the non-rule technological options that might be available as a more effective means of serving regulatory purposes. As a view about the field of legal interest, Law 3.0 highlights the coexistence of the three conversations as well as bringing the use of technological measures for regulatory purposes into the foreground of legal inquiry.

Second, given the tendency of lawyers to default to Law 1.0, it is necessary to reboot legal thinking. Law 1.0 is not the only conversation in which lawyers should engage, and Law 1.0 should not set the limits of legal interest and inquiry. To facilitate this reimagining of law, it is suggested that we should adopt a broad understanding of the regulatory environment (including rules, both formal and informal, and technological measures) as the organising idea within the field of interest.

Third, if the direction of travel is towards a more instrumental way of thinking, as a matter of urgency, we need to retrieve a sense of 'what really matters' for human agents and reconceive the range of regulatory responsibilities accordingly. Contrary to the conventional wisdom, there are foundations (other than tradition or mere recognition or acceptance) for at least some human values and those values can be ranked. Above all, regulators have a stewardship responsibility to protect, preserve, and maintain the global commons, the essential preconditions for human social existence. Then, they have responsibilities for their particular communities, to respect (and treat as privileged) the constitutive values of the particular community and to undertake acceptable balancing of competing and conflicting interests.

Fourth, these elements of a reconceived understanding of the regulatory responsibilities can be cashed out in the form of a triple licence requirement for the use of technological instruments.

Fifth, the authorisation of regulatory action, whether employing rules or technological measures, and whether by public or private regulators, needs to be covered by the Rule of Law. Moreover, the familiar procedural requirements of the Rule of Law need to be connected to substantive provisions that embed the reconceived understanding of regulatory responsibilities and the triple licence. It is imperative that the breadth of the Law 3.0 conversation is complemented by substantive depth so that regulatory thinking is always grounded in, and sensitised to, the deepest infrastructural conditions for human social existence.

Sixth, in place of the formal coherentism of Law 1.0 and the instrumentalist coherentism that is implicit in Law 2.0, Law 3.0 demands a renewed substantive coherentism that checks regulatory action for its compatibility with the range of regulatory responsibilities and the terms of the triple licence.

Seventh, although it is not clear where the Law 3.0 conversation should take place, nor how to translate new coherentism into action, legal institutions (nationally, regionally, and internationally – especially internationally) need to be fit for purpose relative to Law 3.0.

Finally, while legal educators might equate 'thinking like a lawyer' with the mindset of Law 1.0, and while preparing students to operate competently in a Law 1.0 conversation might continue to be focal, legal education should take place against the backdrop given by Law 3.0. In 2020, any legal education that neglects the coexistence of the three conversations and the significance of technology for regulatory purposes needs to be radically overhauled.

No doubt these propositions raise many questions. Here are a few that might be asked (and in some cases have already been), with short responses.

The relationship between law and life

What is the relationship between Law 1.0, 2.0, and 3.0 and Max Tegmark's *Life* 1.0, 2.0, and 3.0 (Tegmark, 2017)? According to Tegmark, we can conceive of life developing through three stages. In Life 1.0, the 'biological' stage, life forms (such as bacteria) simply evolve; in Life 2.0, the 'cultural' stage, which is where we humans find ourselves, biology still develops through evolution but humans are able to learn new skills and develop new tools and technologies; and, in Life 3.0, the 'technological' stage, biology is freed from its evolutionary shackles, life forms now being able, as Tegmark puts it, to design both their hardware and software. In this developmental story, humans instantiate Life 2.0 and the three conversations of Law 1.0, 2.0, and 3.0 all fall within this cultural stage. That said, both narratives highlight the increasing significance of technological development and the overall direction of travel, whether it is to making a life form the master of its own destiny or to increasing the effective control exercised by regulators.

A question about Law 1.0

Is the conversation in Law 1.0 about the application of legal principles, or is it simply about the application of 'the law'? When Law 1.0 is the only conversation in town, it is about the application of the general principles of the common law, as well as about the application of rules that have crystallised in the jurisprudence from these principles, and in this sense it is about the application of the law. However, once the Law 2.0 conversation has formed and now coexists with Law 1.0, legislation that reflects Law 2.0's regulatory thinking comes to be interpreted and applied in the courts. Initially, the coherentist culture of the courts invites a narrow (and literal) interpretation of legislation which is regarded as an exception to the common law. However, as the volume and significance of legislation increases, a more purposive approach is invited. At this point, 'the law' to be applied – comprising both common law principles and rules as well as legislative provisions that serve regulatory policies – reflects an intersection of the thinking that informs two quite different mindsets, Law 1.0 and Law 2.0.

Democracy

How does 'democracy' fit into the picture of Law 3.0? Is it a necessary or a contingent feature of a community that undertakes a Law 3.0 conversation? Although there is more than one conception, and more than one articulation, of democracy, I take it that the core ideas are that regulators are accountable to the community and that the community must be included in regulatory conversations. Accordingly, it follows that members of the community must in some sense be included in the conversation about both the constitutive values of the community and the policies to be adopted by regulators. This implies – in line with my remarks on the Rule of Law in Chapter 19 (as well as about the community and social licences in Chapter 20) – that regulators will not be in a position to discharge their second and third-tier regulatory responsibilities unless they operate in accordance with democracy. It is less clear that democracy is a necessary condition for the discharge of first-tier regulatory responsibilities. However, to the extent that there are difficult choices to be made about which part of the commons to prioritise, democratic debate and accountability are surely necessary.

Enhancing the commons

The stewardship responsibility of regulators has been expressed in terms of protecting, preserving, and maintaining the global commons. However, might there also be a responsibility to *enhance* the conditions of the global commons? If some conditions can be enhanced without any diminution to the other conditions, then this seems to be an easy case; the conditions should be enhanced. But if some conditions can be enhanced only if there is some diminution of other conditions, this looks much more problematic. Possibly, referring back to the previous question, this is a case where there needs to be a democratic debate.

The triple licence

The triple licence is a response to emergent technologies and it is designed to be applicable to the use of technological measures, but is it also applicable to the use of rules? The short answer is that it is.

The practical impact of Law 3.0

How will Law 3.0 impact on practical lawyering? For example, in *Tomorrow's Lawyers*, Richard Susskind (2017) highlights new technologies (particularly information technologies and AI) as one of three key drivers of legal change (the other two being 'more-for-less' or value for money, and the liberalisation of traditionally restrictive legal practices). These drivers suggest that there will be less demand for some traditional legal services because clients will become more adept at servicing their own needs and that the supply of some services will be more automated. Much of this suggests that lawyers will be less heavily involved in servicing those matters that are subject to a Law 1.0 conversation. However, the idea that is implicit in Law 3.0 is that lawyers will play a greater part in other law conversations, whether in regulatory circles or in places where Law 3.0 itself is the conversation. As Susskind emphasises, the future of lawyering 'is not already out there, in some sense pre-articulated and just waiting to unfold' (195). Tomorrow's lawyering is to be made by today's young lawyers. Some of that future lies in a more technological approach to Law 1.0, but it also lies in an engagement with Law 2.0 and, above all, Law 3.0.

The 'imperialism' of Law 3.0

Is the agenda that comes with Law 3.0 'imperialistic'? In one sense it is: it does insist that the field of legal interest should be extended to include both rules and technological measures. It does insist that lawyers should treat both rule-governed situations and situations that are technologically managed as being within their field of interest, and it does insist that Law's Empire extends beyond Law 1.0. In another sense, it is not imperialistic: within the extended field of interest, it accepts that there will be many particular focal points of interest reflected in continuing Law 1.0 and 2.0 conversations. That said, it is part of the agenda of Law 3.0 to intensify and to focus interest on the changing complexion of the regulatory environment and to respond to a range of questions about the legitimacy of technologies being employed with regulatory intent.

Chapter 27

Concluding remarks

Looking back, looking forward

In this book, I have described two ways in which law is disrupted by new technologies. With the first disruption, Law 1.0 is superseded by Law 2.0, and with the second disruption, Law 2.0 is superseded by Law 3.0. In neither case, though, are the earlier conversations and mindsets wholly eclipsed. The state of disruption in which we now find ourselves, Law 3.0, is one of coexistent conversations and mindsets.

Law 3.0 is where we are. To some extent, though, this is an old story. From the industrial revolution onwards, legal rules have needed remedial attention as their deficiencies are exposed – as it becomes apparent that the prevailing rules are not fit for regulatory purpose. That said, the very idea of *a rule not being fit for regulatory purpose* is itself expressive of a radically disrupted way of thinking. Crucially, though, this old story is now joined by a new disruptive story in which two taken-for-granted assumptions are challenged: first, that social ordering is achieved through rules; and, secondly, that the Rule of Law is exclusively about rule by rules. Regulation in the future might be more a matter of a conversation between smart machines than a debate in a legislative forum where the participants are human agents.

Given such disruption, what should we do? I have suggested that we should reframe our thinking, reimagining law as a part of a much broader regulatory environment, an environment that features not only rule-based normative signals but also measures of non-normative technological management. So reimagined, we can develop a jurisprudence that marks up the credentials of rules rather than technological measures and vice versa.

There is no guarantee that rules and technological measures can peacefully coexist. However, if we are to reinvent law, I have suggested that we first need to put in order a grounded and hierarchically ordered scheme of regulatory responsibilities. That scheme, representing the benchmarks of regulatory legitimacy and being cashed out as a triple licence for technological measures, can then be used to inform each community's articulation of the Rule of Law; and it can be taken forward through a new and revitalised form of coherentist thinking together with new institutional arrangements.

Rationally, humans should need little persuading: what we all have in common is a fundamental reliance on a critical infrastructure; if that infrastructure is compromised, the prospects for any kind of legal or regulatory activity, or any kind of persuasive or communicative activity, indeed for any kind of human social activity, are diminished. If we value anything, if we are positively disposed towards anything, we must value the commons. If we cannot agree on that, and if we cannot agree that the fundamental role of law is to ensure that power is exercised only in ways that are compatible with the preservation of the infrastructure of all other infrastructures, then the story of disruption, reimagination, and reinvention certainly will not end well.

That said, a satisfactory accommodation of Law 3.0 might be the end of one chapter but not the end of the story. Technological development and legal disruption will not end with Law 3.0. Looking ahead, there surely will be a Law 4.0.

In *Law, Technology and Society* (2019a), I contemplated the possibility that further developments in human genetics – on the one hand, enabling us to manipulate genes more precisely, and on the other, helping us to understand just how particular genetic profiles account for us being the person that we are, acting and reacting in the way that we do – might encourage a technocratic strategy that focuses on internal coding controls. If so, we would not so much find ourselves in technologically managed environments as be technologically managed through our genetic coding. However, I qualified this by saying that a more attractive option for regulators might be simply to take a much more managed approach to reproduction (compare Greely, 2016).

It might be, though, that biotechnology is not the driver for Law 4.0. Rather, it might be that there are further advances in AI, to the point where humans decide that the entire legal and regulatory enterprise would be better entrusted to the machines. If it is acceptable to have robots staffing prisons, why not also robocops? And, if robocops are acceptable, why not also robojudges? And, if robojudges are acceptable, why not also roboregulators? Of course, there are huge functional leaps in these moves, but a community might be prepared to make them, especially if in other domains, notably health, we entrust life and death decisions and procedures to intelligent machines (compare Tegmark, 2017: 107). In Law 4.0, we might find that it is *Lex Machina* that rules.

But might a community think about doing a technological U-turn? In Law 4.0, might we find that, as in Samuel Butler's *Erewhon* (1872), the machines have been destroyed? To Butler's Victorian readers, living through the transition from Law 1.0 to Law 2.0, it must have seemed quite extraordinary that the Erewhonians – concerned that their machines might develop some kind of 'consciousness,' or capacity to reproduce, or agency, and fearful that machines might one day enslave humans – had decided that the machines must be destroyed. How, Butler's readers must have wondered, could such intelligent and technologically sophisticated people have gone backwards in this way? How, indeed? Perhaps Law 4.0, or subsequent waves of disruption, will shed some light on this particular mystery.

References

Alarie, Benjamin (2016): 'The Path of the Law: Toward Legal Singularity' 66 *University of Toronto Law Journal* 443.

Arnold v Britton (2015): UKSC 36.

Asilomar Conference on Beneficial AI (2017), available at https://en.wikipedia.org/wiki/Asilomar_Conference_on_Beneficial_AI

Barlow, John Perry (1994): 'The Economy of Ideas: Selling Wine without Bottles on the Global Net,' available at www.wired.com/1994/03/economy-ideas/

Bayern, Shawn, Burri, Thomas, Grant, Thomas D., Häusermann, Daniel M., Möslein, Florian, and Williams, Richard (2017): 'Company Law and Autonomous Systems: A Blueprint for Lawyers, Entrepreneurs, and Regulators' 9 *Hastings Science and Technology Law Journal* 135.

Belize Telecom: Attorney General of Belize v Belize Telecom (2009): UKPC 10.

Bridle, James (2018): *New Dark Age: Technology and the End of the Future*, London, Verso.

Brincker, Maria (2017): 'Privacy in Public and Contextual Conditions of Agency,' in Tjerk Timan, Bryce Clayton Newell, and Bert-Jaap Koops (eds), *Privacy in Public Space*, Cheltenham, Edward Elgar, 64.

Brownsword, Roger (2003): 'Human Dignity as the Basis for Genomic Torts' 42 *Washburn Law Journal* 413.

Brownsword, Roger (2014): 'Human Dignity from a Legal Perspective,' in M. Duwell, J. Braarvig, R. Brownsword, and D. Mieth (eds), *Cambridge Handbook of Human Dignity*, Cambridge, Cambridge University Press, 1.

Brownsword, Roger (2017): 'Law, Liberty and Technology,' in Roger Brownsword, Eloise Scotford, and Karen Yeung (eds), *The Oxford Handbook of Law, Regulation and Technology*, Oxford, Oxford University Press, 41.

Brownsword, Roger (2019a): *Law, Technology and Society: Re-Imagining the Regulatory Environment*, Abingdon, Routledge.

Brownsword, Roger (2019b): 'Teaching the Law of Contract in a World of New Transactional Technologies,' in Warren Swain and David Campbell (eds), *Reimagining Contract Law Pedagogy: A New Agenda for Teaching (Legal Pedagogy)*, Abingdon, Routledge, 112.

Brownsword, Roger and Goodwin, Morag (2012): *Law and the Technologies of the Twenty-First Century*, Cambridge, Cambridge University Press.

Brownsword, Roger, Scotford, Eloise, and Yeung, Karen (eds) (2017): *The Oxford Handbook of Law, Regulation and Technology*, Oxford, Oxford University Press.

Brownsword, Roger and Yeung, Karen (eds) (2008): *Regulating Technologies*, Oxford, Hart.

Butler, Samuel (1872): *Erewhon*, Mineola, NY, Dover Thrift (2003).

Bygrave, Lee (2017): 'Hardwiring Privacy,' in Roger Brownsword, Eloise Scotford, and Karen Yeung (eds), *The Oxford Handbook of Law, Regulation and Technology*, Oxford, Oxford University Press, 754.

Calo, Ryan (2014): 'Digital Market Manipulation' 82 *The George Washington Law Review* 995.

Carmarthenshire County Council v Lewis (1955): AC 549.

Cavendish Square Holding BV v Talal El Makdessi (2015): UKSC 67.

Cohen, Julie E. (2019): *Between Truth and Power*, New York, Oxford University Press.

Crootof, Rebecca (2019): '"Cyborg Justice" and the Risk of Technological-Legal Lock-In' 119 *Columbia Law Review* 1.

DC Merwestone (2016): *Versloot Dredging BV and anr v HDI Gerling Industrie Versicherung AG and ors* [2016] UKSC 45.

Durovic, Mateja and Janssen, André (2019): 'Formation of Smart Contracts under Contract Law,' in Larry di Matteo, Michel Cannarsa, and Cristina Poncibò (eds), *The Cambridge Handbook of Smart Contracts, Blockchain Technology and Digital Platforms*, Cambridge, Cambridge University Press, 61.

Easterbrook, Frank H. (1996): 'Cyberspace and the Law of the Horse' *University of Chicago Legal Forum* 207.

Ellul, Jacques (1964): *The Technological Society*, New York, Vintage Books.

European Commission (2015): 'Digital Contracts for Europe: Unleashing the Potential of e-Commerce' COM (2015) 633 final, Brussels.

The Eurymedon (1975): *New Zealand Shipping Co Ltd v AM Satterthwaite and Co Ltd (The Eurymedon)* [1975] AC 154.

Ezrachi, Ariel and Stucke, Maurice (2016): *Virtual Competition*, Cambridge, MA, Harvard University Press.

Fairfield, Joshua (2014): 'Smart Contracts, Bitcoin Bots, and Consumer Protection' 71 *Washington and Lee Law Review Online* 36.

Finck, Michèle (2019): *Blockchain Regulation and Governance in Europe*, Cambridge, Cambridge University Press.

Foer, Franklin (2017): *World without Mind*, London, Jonathan Cape.

Ford, Richard T. (2000): 'Save the Robots: Cyber Profiling and Your So-Called Life' 52 *Stanford Law Review* 1576.

Fox, Dov (2019): *Birth Rights and Wrongs*, New York, Oxford University Press.

Fuller, Lon L. (1969): *The Morality of Law*, New Haven, CT, Yale University Press.

Gal, Michal and Elkin-Koren, Niva (2017): 'Algorithmic Contracts' 30 *Harvard Journal of Law and Technology* 309.

Gash, Tom (2016): *Criminal: The Truth about Why People Do Bad Things*, London, Allen Lane.

Gore, Al (2017): *The Assault on Reason*, updated edn, London, Bloomsbury.

Greely, Henry T. (2016): *The End of Sex and the Future of Human Reproduction*, Cambridge, MA, Harvard University Press.

Guihot, Michael (2019): 'Coherence in Technology Law' 11.2 *Law, Innovation and Technology* 311.

Gunningham, Neil and Grabosky, Peter (1998): *Smart Regulation*, Oxford, Clarendon Press.

Hartzog, Woodrow (2018): *Privacy's Blueprint*, Cambridge, MA, Harvard University Press.

Hawking, Stephen (2018): *Brief Answers to the Big Questions*, London, John Murray.

Heller, Nathan (2016): 'Cashing Out,' *The New Yorker*, October 10, available at www.newyorker.com/magazine/2016/10/10/imagining-a-cashless-world?verso=true

Hildebrandt, Mireille (2008): 'Legal and Technological Normativity: More (and Less) Than Twin Sisters' 12.3 *TECHNE* 169.

Holman v Johnson (1775): 1 Cowp 341.

Hormones (1998): Report of the Appellate Body WT/DS26/AB/R, WT/DS48/AB/R.

House of Lords Select Committee on Artificial Intelligence (2017): *AI in the UK: Ready, Willing and Able?* Report of Session 2017–19, HL Paper 100.

Jasanoff, Sheila (2016): *The Ethics of Invention*, New York, W.W. Norton.

Johnson, David R. and Post, David (1996): 'Law and Borders: The Rise of Law in Cyberspace' 48 *Stanford Law Review* 1367.

Karsales (Harrow) v Wallis (1956): 1 WLR 936.

Katell, Michael, Dechesne, Franciene, Koops, Bert-Jaap, and Meessen, Paulus (2019): 'Seeing the Whole Picture: Visualising Socio-Spatial Power through Augmented Reality' 11.2 *Law, Innovation and Technology* 279.

Kerr, Ian (2010): 'Digital Locks and the Automation of Virtue,' in Michael Geist (ed.), *From 'Radical Extremism' to 'Balanced Copyright': Canadian Copyright and the Digital Agenda*, Toronto, Irwin Law, 247.

Koops, Bert-Jaap (2018): 'Privacy Spaces' 121 *West Virginia Law Review* 611.

LawTech Delivery Panel (UK Jurisdiction Taskforce) (2019): 'Legal Statement on Cryptoassets and Smart Contracts,' available at https://technation.io/about-us/lawtech-panel

Lin, Albert C. (2018): 'Herding Cats: Governing Distributed Innovation' 96 *North Carolina Law Review* 945.

Liu, Hin-Yan (2018): 'The Power Structure of Artificial Intelligence,' 10 *Law, Innovation and Technology* 197.

Llewellyn, Karl N. (1940): 'The Normative, the Legal, and the Law-Jobs: The Problem of Juristic Method' 49 *Yale Law Journal* 1355.

Lucassen, Anneke, Montgomery, Jonathan, and Parker, Michael (2016): 'Ethics and the Social Contract for Genomics in the NHS,' *Annual Report of the Chief Medical Officer 2016: Generation Genome*, available at www.gov.uk/government/uploads/attachment_data/file/624628/CMO_annual_report_generation_genome.pdf

Lukes, Steven (2005): *Power: A Radical View*, 2nd edn, Basingstoke, Palgrave Macmillan.

Macaulay, Stewart (1963): 'Non-Contractual Relations in Business' 28 *American Sociological Review* 55.

Macneil, Ian R. (1980): *The New Social Contract*, New Haven, CT, Yale University Press.

Marchant, Gary E. and Sylvester, Douglas J. (2006): 'Transnational Models for Regulation of Nanotechnology' 34 *Journal of Law, Medicine and Ethics* 714.

Markou, Christopher and Deakin, Simon (2019): 'Ex Machina Lex: The Limits of Legal Computability,' available at https://papers.ssrn.com/sol3/papers.cfm?abstract_id=3407856

Marks, Amber, Bowling, Ben, and Keenan, Colman (2017): 'Automatic Justice? Technology, Crime, and Social Control,' in Roger Brownsword, Eloise Scotford, and Karen Yeung (eds), *The Oxford Handbook of Law, Regulation and Technology*, Oxford, Oxford University Press, 705.

Marks and Spencer plc v BNP Paribas Securities Services Trust Company (Jersey) Limited (2015): UKSC 72.

Mashaw, Jerry L. and Harfst, David L. (1990): *The Struggle for Auto Safety*, Cambridge, MA, Harvard University Press.

Morgan, Jonathan (2017): 'Torts and Technology,' in Roger Brownsword, Eloise Scotford, and Karen Yeung (eds), *The Oxford Handbook of Law, Regulation and Technology*, Oxford, Oxford University Press, 522.

Morozov, Evgeny (2013): *To Save Everything, Click Here*, London, Allen Lane.

O'Malley, Pat (2013): 'The Politics of Mass Preventive Justice,' in Andrew Ashworth, Lucia Zedner, and Patrick Tomlin (eds), *Prevention and the Limits of the Criminal Law*, Oxford, Oxford University Press, 273.

Packer, Herbert L. (1969): *The Limits of the Criminal Sanction*, Stanford, CA, Stanford University Press.

Parker v South Eastern Railway Co (1877): 2 CPD 416.

Pasquale, Frank (2015): *The Black Box Society*, Cambridge, MA, Harvard University Press.

Patel v Mirza (2016): UKSC 42.

Pearl, Tracy (2018): 'Compensation at the Crossroads: Autonomous Vehicles and Alternative Victim Compensation Schemes' 60 *William and Mary Law Review* 1827.

ProCD Inc v Zeidenberg 86 F 3d 1447 (7th Cir. 1996).

Reading Borough Council v Mudassar Ali (2019): EWHC 200 (Admin).

Reynolds, Jesse (2011): 'The Regulation of Climate Engineering' 3 *Law, Innovation and Technology* 113.

Rosenthal, Gert (2017): *Inside the United Nations*, Abingdon, Routledge.

Royal Society and British Academy (2016): *Connecting Debates on the Governance of Data and its Uses*, London, Royal Society and the British Academy.

Rubin, Edward L. (2017): 'From Coherence to Effectiveness,' in Rob van Gestel, Hans W. Micklitz, and Edward L. Rubin (eds), *Rethinking Legal Scholarship*, New York, Cambridge University Press, 310.

Ruxley Electronics and Construction Ltd v Forsyth (1996): AC 344.

S and Marper v The United Kingdom (2009): 48 EHRR 50.

Sayre, Francis (1933): 'Public Welfare Offences' 33 *Columbia Law Review* 55.

Shearing, Clifford and Stenning, Phillip (1985): 'From the Panopticon to Disney World: The Development of Discipline,' in Anthony N. Doob and Edward L. Greenspan (eds), *Perspectives in Criminal Law: Essays in Honour of John LL. J. Edwards*, Toronto, Canada Law Book, 335.

Simpson, Gerry (2004): *Great Powers and Outlaw States*, Cambridge, Cambridge University Press.

State of Wisconsin v Loomis (2016): 881 N.W. 2d 749.

Stena Line v Merchant Navy Ratings Pension Fund Trustees Limited (2011): EWCA Civ 543.

Sunstein, Cass R. (2005): *Laws of Fear*, Cambridge, Cambridge University Press.

Susskind, Richard (2017): *Tomorrow's Lawyers*, 2nd edn, Oxford, Oxford University Press.

Swire, Peter P. and Litan, Robert E. (1998): *None of Your Business*, Washington, D.C., Brookings Institution Press.

Syed, Matthew (2019): 'VAR is Football's Passion Killer: It's Time to Bin It' *The Times*, November 6, 64.

Tegmark, Max (2017): *Life 3.0*, London, Allen Lane.

Thornton v Shoe Lane Parking Ltd (1971): 2 WLR 585.

Tyler, Tom R. (2006): *Why People Obey the Law*, Princeton, NJ, Princeton University Press.

Vallor, Shannon (2016): *Technology and the Virtues*, New York, Oxford University Press.

Viney, Geneviè and Guégan-Lécuyer (2010): 'The Development of Traffic Liability in France,' in Miquel Martin-Casals (ed.), *The Development of Liability in Relation to Technological Change*, Cambridge, Cambridge University Press.

Vos, Geoffrey (2019): 'The Launch of the Legal Statement on the Status of Cryptoassets and Smart Contracts,' available at www.judiciary.uk/announcements/the-chancellor-of-the-high-court-sir-geoffrey-vos-launches-legal-statement-on-the-status-of-cryptoassets-and-smart-contracts/

Walker, Neil (2015): *Intimations of Global Law*, Cambridge, Cambridge University Press.

Wallach, Wendell (2015): *A Dangerous Master*, New York, Basic Books.

Weaver, John Frank (2014): *Robots are People Too*, Santa Barbara, CA, Praeger.

Wood v Capita Insurance Services Ltd (2017): UKSC 24.

Yeung, Karen (2019): 'Regulation by Blockchain: The Emerging Battle for Supremacy between the Code of Law and Code as Law' 82 *Modern Law Review* 207.

Zittrain, Jonathan (2008): *The Future of the Internet*, London, Penguin.

Zuboff, Shoshana (2019): *The Age of Surveillance Capitalism*, London, Profile Books.

Further indicative reading

In the Introduction to this book, I emphasised that the work cited here is simply the tip of an iceberg, that there is now a large and vibrant literature on law, regulation, and technology. This literature is not confined to purely academic books and articles; it includes a large number of 'trade-books' aimed at a general readership as well as all manner of commentaries and posts online.

If we chart the evolution of this literature from the 1990s, we can see that the initial focus was on developments in biotechnology (especially arising from work in genetics) and in information and communications technology (ICT) (especially arising from the work around the Internet). In both cases, there were multiple concerns – for example, GMOs raised concerns about the environment and human health and safety; the possibility of genetic engineering and human cloning gave rise to concerns about discrimination, human rights, and human dignity; and the construction of an embryonic online world gave rise to concerns about equality and digital divides as well as privacy and dignity. And in both cases, there were also major questions about the applicability of proprietary claims (over the human genome and in relation to copyright-protected material) as well as fundamental issues about patentability.

While debates about the regulation of biotechnology and ICT continued to rage, developments in nanotechnology and neurotechnology (particularly the development of fMRI scanning) provoked fresh debates about safety and privacy. This led to a significant extension of the literature. Importantly, in the case of fMRI, while some claimed that this new window into the functioning of the human brain presented fresh opportunities for ascertaining 'the truth' (fMRIs being characterised as a new kind of lie detector), others argued that the findings of neuroscientists raised serious questions about the practice of the criminal justice system in holding offenders responsible for their actions. At much the same time, an interest in the convergence of various elements of bio, nano, neuro, and info-technologies began to take shape, generating a fierce debate about the ethics and governance of human enhancement.

With further developments in additive manufacturing (3D printing) and cryptography (blockchain) as well as major advances in artificial intelligence and machine learning, new questions were presented for regulatory debate – particularly questions about humans being taken out of the legal loop. Moreover, the earlier technologies did not stand still, and new techniques such as gene-editing rekindled earlier debates about safety, autonomy, dignity, and so on.

To a large extent, the central questions in these debates prompted conversations of a Law 1.0 and Law 2.0 kind. However, many of the technologies being developed (such as DNA profiling and various surveillance technologies) shifted the focus from the technologies as, so to speak, regulatory targets to their potential use as regulatory tools. This shift in focus has been further encouraged by recent developments in artificial intelligence and machine learning, all of which presage the formation of a Law 3.0 conversation.

The indicative reading that follows is supplementary to the references already given. In those references, there are some truly seminal publications, but they will not be repeated here. The indicative references here are organised in a way that, broadly speaking, reflects the development of the field.

The law and regulation of biotechnologies

Beyleveld, Deryck and Brownsword, Roger (1993): *Mice, Morality, and Patents (The Oncomouse Application and Article 53(a) of the European Patent Convention)* (with Foreword by Lord Scarman), London, Common Law Institute of Intellectual Property.

Beyleveld, Deryck and Brownsword, Roger (2001): *Human Dignity in Bioethics and Biolaw*, Oxford, Oxford University Press.

Brownsword, Roger (2004): 'Regulating Human Genetics: New Dilemmas for a New Millennium' 12 *Medical Law Review* 14.

Brownsword, Roger (2009): 'Human Dignity, Ethical Pluralism, and the Regulation of Modern Biotechnologies,' in Thérèse Murphy (ed.), *New Technologies and Human Rights*, Oxford, Oxford University Press, 19.

Brownsword, Roger (2016): 'New Genetic Tests, New Research Findings: Do Patients and Participants Have a Right to Know – and do they Have a Right Not to Know?' 8 *Law, Innovation and Technology* 247.

Brownsword, Roger, Cornish, William R., and Llewellyn, Margaret (eds) (1998): *Law and Human Genetics: Regulating a Revolution*, Oxford, Hart Publishing, in conjunction with the *Modern Law Review*.

Brownsword, Roger and Wale, Jeffrey (2018): 'Testing Times Ahead: Non-Invasive Prenatal Testing and the Kind of Community That We Want to Be' 81 *Modern Law Review* 646.

Fukuyama, Francis (2002): *Our Posthuman Future*, London, Profile Books.

Jasanoff, Sheila (2005): *Designs on Nature*, Princeton, NJ, Princeton University Press.

Krimsky, Sheldon and Simoncelli, Tania (2011): *Genetic Justice*, New York, Columbia University Press.

Laurie, Graeme (2002): *Genetic Privacy*, Cambridge, Cambridge University Press.

Lee, Maria (2008): *EU Regulation of GMOs: Law and Decision Making for a New Technology*, Cheltenham, Edward Elgar.

Lee, Maria (2009): 'Beyond Safety? The Broadening Scope of Risk Regulation' 62 *Current Legal Problems* 242.

Pascuzzi, Giovanni, Izzo, Umberto, and Macilotti Matteo (eds) (2013): *Comparative Issues in the Governance of Research Biobanks*, Heidelberg, Springer.

Plomer, Aurora and Torremans, Paul (eds) (2009): *Embryonic Stem Cell Patents: European Law and Ethics*, Oxford, Oxford University Press.
Somsen, Han (ed.) (2007): *The Regulatory Challenge of Biotechnology*, Cheltenham, Edward Elgar.
Widdows, Heather and Mullen, Caroline (eds) (2009): *The Governance of Genetic Information: Who Decides?*, Cambridge, Cambridge University Press.

The law and regulation of ICTs

Boyle, James (1996): *Shamans, Software and Spleens: Law and the Construction of the Information Society*, Cambridge, MA, Harvard University Press.
Edwards, Lilian (ed.) (2017): *Law, Policy and the Internet*, Oxford, Hart.
Goldsmith, Jack and Wu, Tim (2006): *Who Controls the Internet?*, Oxford, Oxford University Press.
Hildebrandt, Mireille and Gutwirth, Serge (eds) (2008): *Profiling the European Citizen*, Dordrecht, Springer.
Kerr, Orin S. (2003): 'The Problem of Perspective in Internet Law' 91 *Georgetown Law Journal* 357.
Koops, Bert-Jaap, Lips, Miriam, Prins, Corien, and Schellekens, Maurice (2006): *Starting Points for ICT Regulation*, The Hague, T.M.C. Asser.
Mayer-Schönberger, Viktor and Cukier, Kenneth (2013): *Big Data*, London, John Murray.
Murray, Andrew (2006): *The Regulation of Cyberspace*, Abingdon, Routledge Cavendish.
Reed, Chris and Murray, Andrew (2018): *Rethinking the Jurisprudence of Cyberspace*, Cheltenham, Edward Elgar.
Reidenberg, Joel R. (2005): 'Technology and Internet Jurisdiction' 153 *University of Pennsylvania Law Review* 1951.

The law and regulation of nanotechnologies

Allhoff, Fritz, Lin, Patrick and Moore, Daniel (2010): *What is Nanotechnology and Why Does It Matter?*, Chichester, John Wiley.
Brownsword, Roger (2008): 'Regulating Nanomedicine: The Smallest of our Concerns?' 2 *Nanoethics* 73.
Hodge, Graeme, Bowman, Diana M., and Maynard, Andrew D. (eds) (2010): *International Handbook on Regulating Nanotechnologies*, Cheltenham, Edward Elgar.
Lin, Albert C. (2007): 'Size Matters: Regulating Nanotechnology' 31 *Harvard Environmental Law Review* 349.
Mandel, Gregory (2008): 'Nanotechnology Governance' 59 *Alabama Law Review* 1323.
Van den Hoven, Jeroen (2007): 'Nanotechnology and Privacy: Instructive Case of RFID,' in Fritz Allhoff, Patrick Lin, James Moor, and John Weckert (eds), *Nanoethics*, Hoboken, NJ, Wiley, 253.

The law and regulation of neurotechnologies

Edwards, Sarah J.L., Richmond, Sarah, and Rees, Geraint (eds) (2012): *I Know What You are Thinking: Brain Imaging and Mental Privacy*, Oxford, Oxford University Press.
Greely, Henry T. (2009): 'Law and the Revolution in Neuroscience: An Early Look at the Field' 42 *Akron Law Review* 687.
Greene, Joshua and Cohen, Jonathan (2004): 'For the Law, Neuroscience Changes Nothing and Everything' 359 *Philosophical Transactions of the Royal Society B: Biological Sciences* 1775.

Illes, Judy (ed.) (2006): *Neuroethics*, Oxford, Oxford University Press.
Morse, Stephen A. (2011): 'Lost in Translation? An Essay on Law and Neuroscience,' in Michael Freeman (ed.), *Law and Neuroscience*, Oxford, Oxford University Press, 529.
Rosen, Jeffrey (2007): 'The Brain on the Stand' *New York Times*, March 11.
Snead, O. Carter (2007): 'Neuroimaging and the "Complexity" of Capital Punishment' 82 *New York University Law Review* 1265.

Convergence

Buchanan, Allen (2011): *Beyond Humanity?*, Oxford, Oxford University Press.
Garreau, Joel (2006): *Radical Evolution*, New York, Broadway Books.
Harris, John (2010): *Enhancing Evolution*, Princeton, NJ, Princeton University Press.
Sandel, Michael J. (2007): *The Case against Perfection*, Cambridge, MA, Harvard University Press.

The law and regulation of additive manufacturing

Li, Phoebe, Faulkner, Alex, and Medcalf, Nicholas (2020): '3D Bioprinting in a 2D Regulatory Landscape: Gaps, Uncertainties, and Problems' 12 *Law, Innovation and Technology* 1.
Mendis, Dinusha, Lemley, Mark, and Rimmer, Matthew (eds) (2019): *3D Printing and Beyond: The Intellectual Property and Legal Implications Surrounding 3D Printing and Emerging Technology*, Cheltenham, Edward Elgar.
Osborn, Lucas S. (2014): 'Regulating Three-Dimensional Printing: The Converging World of Bits and Atoms' 51 *San Diego Law Review* 553.
Tran, Jasper L. (2015): 'To Bioprint or Not to Bioprint' 17 *North Carolina Journal of Law and Technology* 123.

The law and regulation of blockchain (fintech and smart contracts)

Brownsword, Roger (2019): 'Regulatory Fitness: Fintech, Funny Money, and Smart Contracts' 20 *European Business Organization Law Review* 5.
De Filippi, Primavera and Wright, Aaron (2018): *Blockchain and the Law*, Cambridge, MA, Harvard University Press.
di Matteo, Larry A., Cannarsa, Michel, and Poncibò, Cristina (eds) (2019): *The Cambridge Handbook of Smart Contracts, Blockchain Technology and Digital Platforms*, Cambridge, Cambridge University Press.
Hacker, Philipp, Lianos, Ioannis, Dimitropoulos, Georgios, and Eich, Stefan (eds) (2019): *Regulating Blockchain: Techno-Social and Legal Challenges*, Oxford, Oxford University Press.

The law and regulation of artificial intelligence and machine learning

Brownsword, Roger and Harel, Alon (2019): 'Law, Liberty and Technology: Criminal Justice in the Context of Smart Machines' 15 *International Journal of Law in Context* 107.
Hildebrandt, Mireille (2015): *Smart Technologies and the End(s) of Law*, Cheltenham, Edward Elgar.

O'Neil, Cathy (2016): *Weapons of Math Destruction*, New York, Penguin Random House.

Roth, Andrea (2016): 'Trial by Machine' 104 *Georgetown Law Journal* 1245.

Wischmeyer, Thomas and Rademacher, Timo (eds) (2020): *Regulating Artificial Intelligence*, Cham, Springer.

Yeung, Karen and Lodge, Martin (eds) (2019): *Algorithmic Regulation*, Oxford, Oxford University Press.

The use of technologies as regulatory tools

Brownsword, Roger (2005): 'Code, Control, and Choice: Why East is East and West is West' 25 *Legal Studies* 1.

Citron, Danielle Keats (2008): 'Technological Due Process' 85 *Washington University Law Review* 1249.

Larsen, Beatrice von Silva-Tarouca (2011): *Setting the Watch: Privacy and the Ethics of CCTV Surveillance*, Oxford, Hart.

Lessig, Lawrence (1999a): *Code and Other Laws of Cyberspace*, New York, Basic Books.

Lessig, Lawrence (1999b): 'The Law of the Horse: What Cyberlaw Might Teach' 113 *Harvard Law Review* 501.

Lessig, Lawrence (2006): *Code 2.0*, New York, Basic Books.

Reidenberg, Joel R. (1997): 'Lex Informatica: The Formulation of Information Policy Rules Through Technology' 76 *Texas Law Review* 553.

Tien, Lee (2004): 'Architectural Regulation and the Evolution of Social Norms' 7 *Yale Journal of Law and Technology* 1.

Yeung, Karen (2011): 'Can We Employ Design-Based Regulation While Avoiding *Brave New World*?' 3 *Law, Innovation and Technology* 1.

Printed in the United States
by Baker & Taylor Publisher Services